Sustainable Consumption and Production

Editors

Dr. Mamta Sharma   Dr. Hukam Singh

Dr. Upendra Singh

Pustak Bharati
Toronto Canada

Editors : Dr. Mamta Sharma
Dr. Hukam Singh
Dr. Upendra Singh

Book Title : Sustainable Consumption and Production

Cover Picture : By Dr. Anil Kumar Chhangani, D. Sc.

Published by :
Pustak Bharati (Books India)
180 Torresdale Ave, Toronto Canada M2R3E4
email : pustak.bharati.canada@gmail.com
Web : www.pustak-bharati-canada.com

Published for
Raj Rishi Government Autonomous College,
Alwar, Rajasthan, India

Financial Assistance
Rashtriya Uchchatar Shiksha Abhiyan
(RUSA-2.0)

Preface

"Our planet is slowly dying, and if we don't do anything about it soon enough, it would eventually begin to deteriorate and everything would be used. The world would become a barren place without any resources. We need to cater to the needs of our planet, and we need to change our life styles so that it becomes beneficial to the planet. We need to become much more eco-friendly, so that no harm is dealt to the planet by our existence. Many people don't realize that they waste large amounts of energy and other resources in various unnecessary things that could otherwise be saved."

This series of books is an extension of the 3 days international conference on **Multidisciplinary Approach Towards Sustainable Development and Climate Change for A Viable Future (ICMSDC-2022)** held from 12th -14th August 2022 at Raj Rishi Government Autonomous College, Alwar, Rajasthan.

We are very happy and delighted to publish our series of books which are accumulation of research papers of knowledgeable experts in the field of sustainable development and climate change.

Climate change is the most significant challenge to achieving sustainable development, and it threatens to drag millions of people into grinding poverty. At the same time, we have never had better know-how and solutions available to avert the crisis and create opportunities for a better life for people all over the world. Climate change is not just a long-term issue. It is happening today, and it entails uncertainties for policy makers trying to shape the future.

There is a dual relationship between sustainable development and climate change. On the one hand, climate change influences key natural and human living conditions and thereby also the basis for social and economic development, while on the other hand, society's priorities on sustainable development influence both the greenhouse gas emissions that are causing climate change and the vulnerability.

Climate policies can be more effective when consistently embedded within broader strategies designed to make national and regional development paths more sustainable. This occurs because the impact of climate variability and change, climate policy responses, and associated socio-economic development will affect the ability of countries to achieve sustainable development goals. Conversely, the pursuit of those goals will in turn affect the opportunities for, and success of, climate policies.

With these books, we aim to reach to as many people as we can, and spread awareness about sustainable development and climate change and its in-depth analysis through our didactic research papers. We hope that the thought with which ICMSDC-2022 was executed is taken forward through this series of books and the inception of an idea of saving the environment is rooted in the minds of our readers. The articles in these books have been contributed by eminent research scholars, scientists, academicians and industry experts whose contributions have enriched this book series. We thank our publisher, Pustak Bharati, Toronto, Canada for joining us in this initiative and helped in publishing this series of books.

Finally, we will always remain indebted to all our well-wishers for their blessings, without which ICMSDC-2022 and series of these book would have not come into existence.

Financial Assistance provided by Rashtriya Uchchatar Shiksha Abhiyan (RUSA-2.0) is gratefully acknowledged.

Dr. Mamta Sharma
Dr. Hukam Singh
Dr. Upendra Singh

Contents

1. Energy Conservation and Environment

Dr. Mamta Sharma*
Dr. Hukam Singh**
Dr. Upendra Singh ***

Introduction

Energy conservation is not about making limited resources last as long as they can, that would mean that you are doing nothing more than prolong a crisis until you finally run out of energy resources altogether. Conservation is the process of reducing demand on a limited supply and enabling that supply to begin to rebuild itself. Many times, the best way of doing this is to replace the energy used with an alternate source. The few energy conservation techniques may surprise us. While there are practical methods such as insulation, changing light sources, using alternate fuels and carpooling rather than walking – understand the 7 core techniques beneath that will show you more about what to do in life. The goal with energy conservation techniques is to reduce demand, protect and replenish supplies, develop and use alternative energy sources, and to clean up the damage from the prior energy processes.

Detour :

Energy conservation is when we make a conscious decision to use less energy. When we lower the amount of energy we use, we slow down fossil fuel depletion and help clean the environment. The top benefits of conserving energy are – (i) help the environment, (ii) prolong the existence of fossil fuels, (iii) save money.

When we limit our energy usage, we can decrease the impact on the environment. The longer we go without making significant changes, the greater the threat of global warming and climate change become to our daily lives.

When we burn fossil fuels, they create incredible amount of greenhouse gas emissions. These gases, which include CO_2, add up faster than the atmosphere can absorb them which prevents Earth from being able to maintain a stable temperature properly. And the side effects are classified as climate change. This brings many harmful effects such as changes in sea level, cold snaps, droughts, hurricanes, melting glaciers and wildfires.

Fortunately, we can reduce greenhouse gas emissions by reducing our use of energy.

The simple act of energy conservation can help slow global warming which allow us to:

- Save coastal cities from disappearing under water
- Improve water quality and protect reefs and other fragile ecosystems
- Improve air quality and reduce airborne allergies leading to a reduced risk of cardiovascular and respiratory issues
- Lower impact on mental health injuries and fatalities caused by severe weather

When we reduce our energy usage, we are playing an important role in helping the environment. Energy efficiency is a critical component in curbing climate change and protecting our environment.

Importance, Benefits and Ways to save the Energy :

Energy conservation is the effort to reduce wasteful energy consumption by using fewer energy services. This can be done by using energy more effectively (using less energy for continuous service) or changing one's behaviour to use less service (for example, by driving less).

Energy conservation can be broadly defined as the efficient use of energy. The reduction or elimination of unnecessary or unwanted energy use is referred to as energy conservation. It can be accomplished by using less energy to perform a given amount of work or by not using energy at all. Globally, energy conservation has enormous potential to boost economic growth while reducing **greenhouse gas emissions** (GHG). More than two decades ago, India began implementing energy conservation measures and has since established a clear policy architecture for promoting energy conservation.

It can be attained through efficient energy use, in which energy use is reduced while achieving the same result, or through reduced consumption of energy services. It is one of the simplest ways to help the environment by reducing pollution and utilising natural energy.

Importance of Energy Conservation

Energy conservation is an important factor in mitigating climate change. It aids in the substitution of renewable energy for non-renewable resources. **Energy conservation** is frequently the most cost-effective solution to energy shortages, as well as a more environmentally friendly alternative to increased energy production.

The Importance of Energy Conservation is as follows :

1. When we save energy, we save the country a lot of money - Approximately 75% of our crude oil requirements are met through imports, which cost approximately Rs.1,50,000 crore per year.
2. We save money when we save energy. Consider how much money you could save if your LPG cylinder arrived for an extra week or if your electricity bills were reduced.
3. The majority of the energy sources we use cannot be reused or renewed – non-renewable energy sources account for 80% of total fuel consumption. It is estimated that our energy resources will only last another 40 years or so.
4. We consume energy faster than it can be produced. The most commonly used energy sources, coal, oil, and natural gas, take thousands of years to form.
5. We save energy by saving energy - When we use fuel wood efficiently, our fuel wood requirements are reduced, as is our drudgery for collecting it.
6. Energy saved equals energy generated - Saving one unit of energy equals producing two units of energy.
7. Save energy to reduce pollution - Energy production and consumption account for a significant portion of air pollution and more than 83 percent of greenhouse gas emissions.
8. Energy resources are limited - India has about 1% of the world's energy resources but 16% of the world's population.

Benefits of Energy Consumption

1. **Promote Health :** Pollution causes or worsens a wide range of serious medical issues, including lung cancer and asthma. By conserving fuel, people can also protect the health of their fellow humans.

2. **Reduce Living Expenses :** When the general public consumes less fuel or electricity, prices fall as a result.

3. **Limited Resources :** People can focus on ensuring that budget - friendly energy is available for future generations by conserving electricity.

4. **Benefits Environment as well as Protect Wildlife :** It reduces direct air pollution caused by machinery, vehicles, and power plants. Conservation also lowers the number of hazardous extraction projects and spills. It can make better environmental outcomes.

5. **Fewer Power Plants :** Utilities will not need to build as many power plants if the public conserves electricity.

6. **Reduce Dependence :** Conservation makes relying on local and regional energy supplies more feasible.

Ways of Energy Conservation :

The primary way of energy conservation is to use clean and alternative sources of energy like wind energy, solar energy, tidal energy, and biomass energy. We can reduce the use of fossil fuels like coal, petroleum, and natural gas by switching to these energy sources. These energy sources are abundant in nature and can be harnessed at any time in any amount. Moreover, they are cheaper than fossil fuels.

Some other ways of Energy Conservation are as follows :

a. We should use CFL bulbs and LEDs instead of regular incandescent bulbs. Using CFLs will reduce the cost per unit of energy consumed. On the other hand, LEDs consume less energy than a regular incandescent bulb

b. We should buy star-rated electrical appliances. An electrical device with a higher number of stars will consume less energy and reduces the cost per unit of energy.

c. We should use sunlight in our homes, schools, and workplaces during the daytime. This will reduce our electricity bills.

d. We should switch off fans, lights, and other electronic devices when not in use or before going out of the room.

e. We should reduce the use of vehicles for going to places that are within walking distance, and we can increase the use of bicycles. This will reduce the unnecessary consumption of fuels like petroleum and CNG. It also helps in reducing air pollution.

f. We must keep the windows and doors of our rooms open to ventilate them with natural air instead of using exhaust fans.

g. We must use devices that work with the thermostat. It will automatically turn off devices when the desired temperature level is achieved. For example, it is used in geysers. This will reduce electricity consumption by devices when not required.

From this article, we can conclude that we cannot create or destroy energy. But we can transform one form of energy into another form. There are some limited resources of energy that must be conserved for present and future generations. We should use energy-efficient devices and minimize the use of non-renewable resources in order to achieve a good quality of life and a clean environment.

Practical Methods of Energy Conservation

Below are 15 energy conservation techniques that can help you to reduce your overall carbon footprint and save money in the long run.

1. Install CFL Lights

Try replacing incandescent bulbs in your home with CFL bulbs. CFL bulbs cost more upfront but last 12 times longer than regular incandescent bulbs. CFL bulbs will not only save energy but over time you end up saving money.

2. Lower the Room Temperature

Even a slight decrease in room temperature, let's say by only a degree or two, can result in big energy savings. The more the difference between indoor and outdoor temperature, the more energy it consumes to maintain room temperature. A smarter and more comfortable way of doing this is to buy a programmable thermostat.

3. Fix Air Leaks

Proper insulation will fix air leaks that could be costing you more. During winter months, you could be letting out a lot of heat if you do not have proper insulation. You can fix those leaks yourself or call an energy expert to do it for you.

4. Use Maximum Daylight

Turn off lights during the day and use daylight as much as possible. This will reduce the burden on the local power grid and save you a good amount of money in the long run.

5. Get Energy Audit Done

A home energy audit is nothing but a process that helps you to identify areas in your home where it is losing energy and what steps you can take to overcome them. Implement the tips and suggestions given by those energy experts, and you might see some drop in your monthly electricity bill.

6. Use Energy Efficient Appliances

When planning to buy some electrical appliances, prefer to buy one with Energy Star rating. Energy-efficient appliances with Energy Star rating consume less energy and save you money. They might cost you more in the beginning, but it is much more of an investment for you.

7. Drive Less, Walk More and Carpooling

Yet another energy conservation technique is to drive less and walk more. This will not only reduce your carbon footprint but will also keep you healthy as walking is a good exercise. If you go to the office by car and many of your colleagues stay nearby, try doing carpooling with them. This will not only bring down your monthly bill you spend on fuel but will also make you socially more active.

8. Switch off Appliances When Not in Use

Electrical appliances like coffee machines, idle printers, desktop computers keep on using electricity even when not in use. Just switch them off if you don't need them immediately.

9. Plant Shady Landscaping

Shady landscaping outside your home will protect it from intense heat during hot and sunny days and chilly winds during the winter season. This will keep your home cool during the summer season and will eventually turn to big savings when you calculate the amount of energy saved at the end of the year.

10. Install Energy Efficient Windows

Some of the older windows installed at our homes aren't energy efficient. Double panel windows and other vinyl frames are much better than single-pane windows. Choosing correct blinds can save on your power bills.

11. Bicycles are your best friend

Yes, bicycles could help us a lot in the process of energy conservation. Since the bicycles are manually driven and use no forms of energy whatsoever, but manpower, the bicycles are literally your best friend. If you are health conscious, then it is simply double the gain.

Just by choosing to go from one place to the other by means of bicycles, you are not only doing your bit towards saving the environment but also you are striving towards a healthier lifestyle.

Altogether, it is good for both you and for the purposes of energy conservation. Keeping all this in view in mind, many countries across the world have designed specialized bicycle lanes to promote the use of the bicycle and also to promote the safety of the ones riding it.

12. Buying a Programmable Thermostat is the best decision to make

In our life, we all have had that moment where we have known that we do not need the air conditioner or the heater, but we have been lazy enough to ignore that inner voice. With a programmable thermostat, life just becomes a tad bit more efficient.

A programmable thermostat can turn itself on and off as and when required without troubling you, even when you are away. It is not just cost-saving, but it is also something that would help you do your bit towards energy conservation.

13. Motion Detectors are a Real Saviour

Installing motion detectors could help you a lot in serving the purpose of energy conservation, besides cutting down your energy bill budgets. Installing motion detectors, especially for external lighting, could be really helpful.

This way, you do not have to worry about turning them off when you are leaving or even worry about turning them on when you get back, which means, you no longer have to turn the key in absolute darkness. Motion detectors would make your life a little bit more convenient.

14. Closing doors is the key to conserving energy

It cannot get any more literal than this. Shutting the doors immediately behind you is one of the best ways to conserve energy. This includes the refrigerator doors as well as the doors of the rooms where an air conditioner is running. This saves not only the machines but also your electricity bills. Also, above all, it is your little step towards the conservation of energy.

15. Keep Your Dryers as Clean as Possible

Cleaning out the lint filter of your dryer could save a lot of energy as well.

References :
Source of knowledge is the internet and it is highly acknowledged.

*Associate Professor (Zoology)
**Professor
*** Associate Professor (Chemistry)
Raj Rishi Government (Autonomous) College
Alwar, Rajasthan 301001,India.
email : mamta810@gmail.com
drhukamsingh63@gmail.com
dr.usingh09@gmail.com

2. A Study of the Impact of Urbanization on Biodiversity

Mr. Vikas Pawar [1]
Mr. Dattatray Katore [2]
Mrs. Chhaya Vanjare [3]

Abstract

The purpose of this study is to look into how urbanisation affects biodiversity. This study aims to investigate how much urbanisation has impacted biodiversity in various regions of the world. Urbanization is recognised to have detrimental consequences on the environment. According to the study, urbanisation significantly reduces biodiversity, with urban regions having lower biodiversity levels than rural ones. The study emphasizes the importance of sustainable infrastructure, green areas, and environmentally friendly structures in order to mitigate the detrimental effects of urbanisation on biodiversity. Its impact was mostly caused by changes in land use, pollution, and climate change. However, the study also identified potential mitigation measures, such as the preservation of green spaces, sustainable urban planning, and public awareness and education campaigns, to lessen the adverse effects of urbanisation on biodiversity. Policymakers, urban planners, and local communities should be aware of the implications of this study's findings, which underscore the need for efficient ways to preserve biodiversity in the face of accelerating urbanisation. The results of this study have significant ramifications for urban planning, emphasizing the necessity of incorporating biodiversity conservation into plans for urban growth.

Keywords : Biodiversity, Ecology, Conservation, Urbanisation, Sustainable

Introduction

Biodiversity

The term "biodiversity" refers to the range of living things that exist on Earth, such as plants, animals, fungi, and microorganisms, as well as the ecological systems and processes that sustain them. A vital element of the natural world is biodiversity, which offers crucial ecosystem services like air and water filtration, nutrient cycling, and climate management.

For the survival and well-being of all living things, including humans, biodiversity is crucial. It aids in the production of essential resources such as food, medicine, and other essentials and enhances ecosystems' resilience and adaptability to environmental upheaval and change.

Even so, widespread declines in biodiversity are being brought on by human activities like habitat destruction, pollution, overuse of natural resources, and climate change, which could have negative effects on ecosystems and human societies. Thus, preserving and protecting biodiversity is essential for sustainable development and the welfare of future generations.

Objective of Study

- To assess the level of urbanization and how it affects biodiversity in the research area.
- To determine the main causes of the urbanization-related loss of biodiversity.
- To identify the key participants and the functions they perform in urban biodiversity conservation.

Types of Biodiversity

Genetic Diversity

The term "genetic diversity" refers to the range of genes and genetic traits present in a species. The ability of species to adapt and evolve in response to shifting environmental conditions is made possible by genetic variety. A population may be more susceptible to illness, predators, or other dangers without genetic diversity.

Species Diversity

Species diversity is a sort of biodiversity that describes the diversity of species that can be found within an ecosystem or throughout the entire world. Because it provides crucial ecological functions including pollination, nutrient cycling, and pest control, species diversity is significant. Moreover, it aids in the resilience and stability of ecosystems.

Ecosystem Diversity

Ecosystem diversity: This category of biodiversity relates to the diversity of many ecosystems found on Earth, including forests, grasslands, wetlands, and seas. Ecosystem diversity is crucial because it creates habitat for a variety of species and contributes to the stability of the biogeochemical cycles on Earth.

The overall well-being and efficiency of the planet's ecosystems are influenced by and dependent upon all three categories of biodiversity. The health of both people and the natural environment depends on maintaining biodiversity.

Urbanization

Urbanization is the process of a growing population concentration in cities or urban areas. This pattern has been present across the world for many years, and it is anticipated to continue in the years to come.

Economic possibilities, greater infrastructure and services, and better living conditions are some of the main drivers of urbanization. Demand for housing, transportation, employment opportunities, and other facilities increases as more people move to metropolitan areas, which in turn stimulates economic growth and development.

Yet, urbanization also brings with it problems like traffic jams, pollution, substandard housing, and socioeconomic inequality. If not properly addressed, these issues could result in social, economic, and environmental issues.

Governments and other stakeholders must establish policies and strategies that support sustainable urban development to overcome the problems caused by urbanization. This covers the availability of cheap housing, effective public transportation, accessibility to essential services, and environmental protection. Communities' needs and priorities can also be considered by incorporating them in the planning and decision-making processes.

Reason for Urbanization

Urbanization is the process through which people move from rural to urban areas, causing the expansion of cities and towns. Urbanization has occurred and is now occurring for a number of reasons worldwide:

Economic Opportunities : People relocate to metropolitan areas in quest of better career chances since cities offer more job options and greater incomes than rural locations.

Access to Services : Those looking for a higher quality of life are drawn to cities because they provide easier access to healthcare, education, transportation, and other necessary services.

Technological Developments : Access to infrastructure that might enhance people's life, such as high-speed internet, is frequently better in urban regions.

Social and Cultural Aspects : Those seeking new experiences may find urban places intriguing since they offer a wider diversity of people, cultures, and lifestyles than rural areas.

Greater Quality of Life : With access to entertainment, cultural events, and social networks, urban locations provide a better quality of life. This appeals especially to younger generations who are looking for a more active way of life.

Thus, even while urbanization has many advantages, it is crucial to take steps to lessen its adverse effects on the environment and to take into account how it may affect biodiversity. This can be done by prioritizing biodiversity protection and natural habitat preservation in urban planning and development initiatives.

1. More than 2/3rd of the world will be urban in 2050

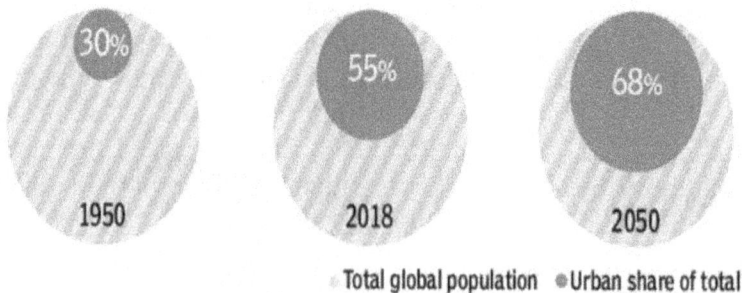

30%   55%   68%

1950   2018   2050

Total global population ● Urban share of total

TOI FOR MORE INFOGRAPHICS DOWNLOAD TIMES OF INDIA APP App Store Google play Windows Phone

Source-https://timesofindia.indiatimes.com/india/7-charts-show-why-india-is-a-world-leader-in urbanisation/ articleshow/ 70325128.cms

2. India will lead the world in urban growth

% of global urban growth ▼

Rank	Growth in urban population (2018-2050)	
#1 **India**	416 million people	17
#2 China	255 million people	10
#3 Nigeria	189 million people	8

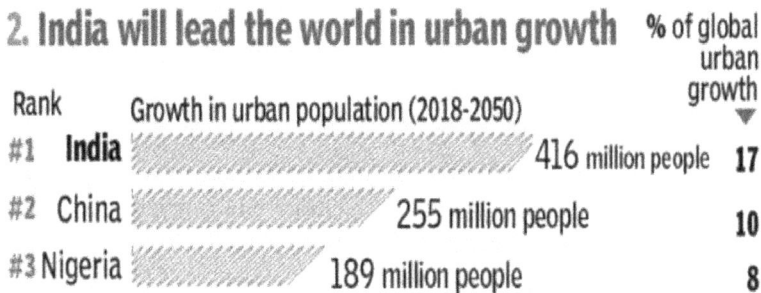

Just India, China and Nigeria will account for 35% of global urban population growth

TOI FOR MORE INFOGRAPHICS DOWNLOAD TIMES OF INDIA APP App Store Google play Windows Phone

Source-https://timesofindia.indiatimes.com/india/7-charts-show-why-india-is-a-world-leader-in urbanisation/ articleshow/70325128.cms

Causes of The Urbanization-Related Loss of Biodiversity

Habitat Destruction : Urbanization frequently results in the transformation of natural habitats into urban landscapes, including marshes, grasslands, and woods. This limits the amount of area accessible for wildlife and ruins the habitats of numerous species.

Habitat Fragmentation : The breaking up of natural habitats into smaller, isolated parts is another effect of urbanization. This may restrict species' freedom of movement and access to resources, which may result in a decline in biodiversity.

Pollution : Urban environments produce a lot of pollution, including light, noise, and environmental pollutants in the air and water. Animal health and survival may suffer because of this.

Introduction of Non-native Species : Urbanization may mistakenly or intentionally result in the introduction of non-native species. These species have the potential to destabilize ecosystems, outcompete native species for resources, and cause biodiversity loss.

Climate Change : Urbanization can have an impact on climate change, which can have a large negative impact on biodiversity. For instance, changes in temperature and precipitation patterns can affect how many species are present and where they are found.

Human Activities : Human populations and their pursuits, such as fishing, trapping, and hunting, expand together with the expansion of urban centers. These actions may have an immediate effect on animal populations and lead to biodiversity loss.

Biodiversity Conservation

The protection, management, and sustainable use of the natural resources on Earth, such as the wide variety of plant and animal species, ecosystems, and genetic diversity, are together referred to as biodiversity conservation. Biodiversity has various ecological, economic, cultural, and social advantages for people and is crucial for sustaining the harmony of nature.

Protecting species and their habitats, restoring degraded ecosystems, and promoting sustainable development strategies that strike a balance between economic expansion and environmental protection are all part of conservation initiatives. The preservation of biodiversity involves all three spheres of society: governments, non-governmental organizations, and individuals.

Understanding the natural mechanisms that support biodiversity as well as the social, economic, and cultural elements that affect conservation efforts is essential for effective conservation. The creation of protected areas, the adoption of sustainable land-use techniques, and the encouragement of public awareness and education campaigns are all examples of conservation activities.

The long-term survival of our world and the welfare of future generations depend on the conservation of the biodiversity. We can preserve the intricate web of life that ensures our survival and the survival of countless other species on Earth by maintaining biodiversity.

Sustainable Consumption and Production

3. Most Indians will live in urban areas before 2050
Percentage of population in urban and rural areas

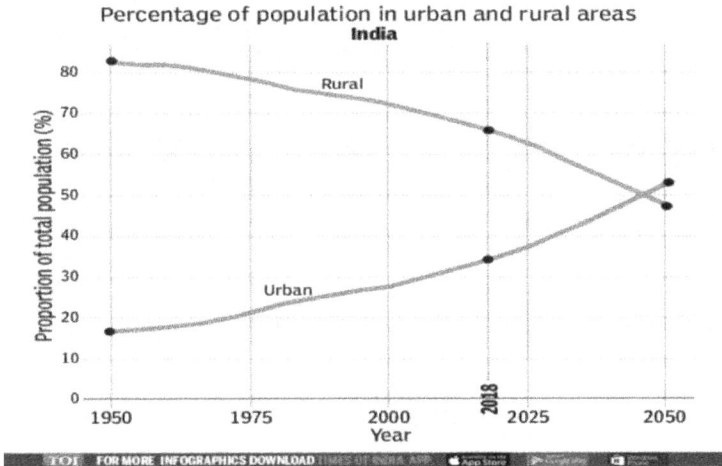

India

Source- https://timesofindia.indiatimes.com/india/7-charts-show-why-india-is-a-world-leader-in-urbanisation/ articleshow/70325128.cms

Steps for Biodiversity Conservation

The practice of defending and preserving the variety of life on Earth is known as biodiversity conservation. The following actions can be performed to preserve biodiversity:

List the environments and species that need to be protected: Find out which species and habitats are most vulnerable to extinction or deterioration by conducting study on them.

Create Protected Areas : Create protected areas to preserve the ecosystems that support endangered species and to offer habitat for those species.

Control Invasive Species : Prevent invasive species from posing a harm to native wildlife or plants by displacing them or bringing in new illnesses.

Reduce Pollution : Decrease pollution and the use of dangerous chemicals that can affect ecosystems and wildlife.

Promote Eco-tourism and Sustainable Development : Promote eco-tourism and sustainable agriculture as examples of sustainable development strategies that strike a balance between economic expansion and environmental protection.

15

Promote Local Communities : Participate in local decision-making, give them knowledge and training, and provide incentives for conservation to involve local communities in efforts to conserve biodiversity.

Conclusion

The research on how urbanization affects biodiversity reveals the considerable and harmful implications that accelerated urbanization can have on the ecosystem. According to research, urbanization causes species diversity and abundance to decrease, habitats to become fragmented, and ecosystem functions to change.

These detrimental effects are a result of a number of factors, such as increased land use change, habitat destruction, and pollution. In addition, urbanization frequently results in the introduction of invasive and non-native species, which further disturbs the ecosystems' natural equilibrium.

The study emphasizes the value of protecting urban biodiversity hotspots, green spaces, and natural habitats in order to lessen the adverse effects of urbanization. Urban areas' ecological value can be increased, and biodiversity can be preserved by incorporating sustainable practices into urbanization planning and implementation.

References

Li, X., Stringer, L. C., & Dallimer, M. (2022). The Impacts of Urbanisation and Climate Change on the Urban Thermal Environment in Africa. In *Climate* (Vol. 10, Issue 11). MDPI. https://doi.org/10.3390/cli10110164

Bhuvandas, N., & Vallabhbhai, S. (2012). *Impacts of urbanisation on environment Analysis of trend of Extreme Daily Temperature of Abu Dhabi city, UAE View project.* https://www.researchgate.net/publication/265216682

Dr BAnandapriya Dr BJeeva Rekha DrAAngel cerli, E., & Subhash Kadam Dr Rajkumari, P. (n.d.). *Digital Transform Using Emerging Technologies.*

Kondratyeva, A., Knapp, S., Durka, W., Kühn, I., Vallet, J., Machon, N., Martin, G., Motard, E., Grandcolas, P., & Pavoine, S.

(2020). Urbanization Effects on Biodiversity Revealed by a Two-Scale Analysis of Species Functional Uniqueness vs. Redundancy. *Frontiers in Ecology and Evolution, 8.* https://doi.org/10.3389/fevo.2020.00073

Simkin, R. D., Seto, K. C., Mcdonald, R. I., & Jetz, W. (n.d.). *Biodiversity impacts and conservation implications of urban land expansion projected to 2050.* https://doi.org/10.1073/pnas.2117297119/-/DCSupplemental

Zimmerer, K. S., Duvall, C. S., Jaenicke, E. C., Minaker, L. M., Reardon, T., & Seto, K. C. (2021). Urbanization and agrobiodiversity: Leveraging a key nexus for sustainable development. In *One Earth* (Vol. 4, Issue 11, pp. 1557–1568). Cell Press. https://doi.org/10.1016/j.oneear.2021.10.012

de Albuquerque, F. S., Bateman, H. L., Boehme, C., Allen, D. C., & Cayuela, L. (2021). Variation in temperature, precipitation, and vegetation greenness drive changes in seasonal variation of avian diversity in an urban desert landscape. *Land, 10*(5). https://doi.org/10.3390/land10050480

Buczkowski, G., & Richmond, D. S. (2012). The effect of urbanization on ant abundance and diversity: A temporal examination of factors affecting biodiversity. *PLoS ONE, 7*(8). https://doi.org/10.1371/journal.pone.0041729

Concepción, E. D., Moretti, M., Altermatt, F., Nobis, M. P., & Obrist, M. K. (2015). Impacts of urbanisation on biodiversity: The role of species mobility, degree of specialisation and spatial scale. *Oikos, 124*(12), 1571–1582. https://doi.org/10.1111/oik.02166

[1]**Assistant Professor,**
[2]**Assistant Professor,**
3 Assistant Professor,
Dr. D. Y. Patil Vidyapeeth,
Center for Online Learning, Pune, Maharashtra, India,
email : vikas.pawar.col@dpu.edu.in
dattatray.katore.@dpu.edu.in
Chhaya.vanjare.col@dpu.edu.in

3. A Brief Analysis of the Environmental Effect of COVID-19 Pandemic

Lata, Rajendra*

Abstract

The existing COVID-19 pandemic has significantly affected numerous facets of daily life, causing a sharp decline in industrialization, transportation on the roads, and ecotourism in a relatively brief period of time. On March 13, 2020, the unprecedented coronavirus disease (COVID-19) was proclaimed a pandemic, and as a result of its explosive growth, it has become a major worldwide emergency. There have been no reports of antiviral medications or immunizations that are therapeutically successful against COVID-19 up to this point. A critical evaluation of research exploring the impact of COVID-19 by and on environmental parameters have been discussed. This study gives a summary of the reported harm done to the environment and society, and it also discusses potential controls for the disease.

Graphical Abstract :

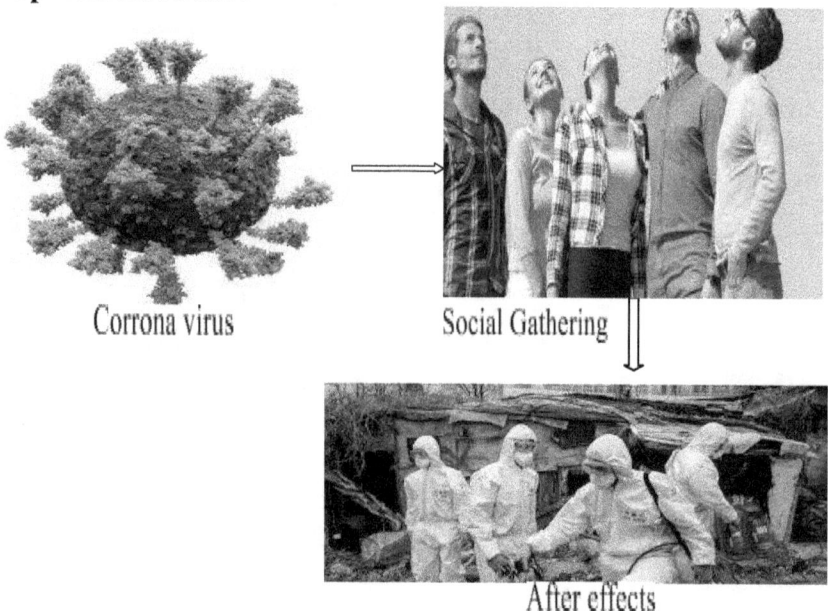

Corrona virus

Social Gathering

After effects

Keywords : COVID-19, Pandemic situation, Lockdown, Pollution

Introduction

Infections of the respiratory tract in people can vary in severity from the ordinary cold to more serious conditions like the Middle East Respiratory Syndrome (MERS) and Severe Acute Respiratory Syndrome, which are all known to be caused by different viruses (SARS). COVID-19 is the name given by the World Health Organization (WHO) to a new infectious respiratory disorder which first emerged in Wuhan, Hubei Province, China, in December 2019. A category of viruses known as coronaviruses (CoVs) infect humans via virus transmission. Covid-19 are 65 nm to 125 nm in diameter and have single-stranded RNA as their nuclear material. The virus's quick propagation and protracted incubation time are by far its most noticeable characteristics. Covid-19's worldwide scale spread, repetitive emergence, substantial amount of fatalities, contamination and death among care workers, and multiplicative impact in weak or vulnerable groups are all important reasons of worry. Covid-

19 was designated a pandemic illness by the Director General of the World Health Organization on March 11th, and he also informed on the 13-fold rise in positive cases in China and 114 nations suffering from 1, 18,000 optimistic cases and 4291 losses to date. It is known that the virus has no direct effect on the climate or the energy industry. However, societal circumstances may emerge that have an indirect impact on the atmosphere and energy industry. The consequences of about effective and conclusive steps opted by nations to avoid Covid-19 are only now becoming apparent. The impacts of the Covid-19 breakout on human lifecycle have begun to be examined from various perspectives, and it was discovered that 35 million COVID-19 cases and over 1,034,000 related deaths had been recorded globally as of October 4th. With the rise in COVID-19 cases, the hunt for effective vaccines and medicines is intensifying, and they are currently in clinical trials.[1-3]

Sustainable Consumption and Production

S. No.	Drug Name	CAS No	Molecular Weight	Formula	Image	Side effects
1.	Nelfinavir	159989-64-7	663.9	$C_{32}H_{45}N_3O_4S$	**Viracept** Nelfinavir	✓ Diarrhea ✓ Nausea or gas ✓ headaches ✓ joint pain ✓ vision changes ✓ overactive thyroid ✓ Guillain-Barre syndrome ✓ blood sugar rise ✓ allergic reaction
2.	Lopinavir	192725-17-0	628.814	$C_{37}H_{48}N_4O_5$		✓ trouble breathing ✓ cough ✓ Fever ✓ bulging eyes ✓ goiter ✓ dark urine,
3.	Remdesivir	1809249-37-3	602.6	$C_{27}H_{35}N_6O_8P$		✓ Back pain ✓ chest tightness ✓ chills ✓ cough ✓ dark-colored urine ✓ trouble breathing ✓ unusual tiredness or weakness ✓ yellow eyes or skin
4.	Favipiravir	259793-96-9	157.104	$C_5H_4FN_3O_2$		✓ Diarrhoea ✓ Hyperuricemia (elevated uric acid) ✓ Reduced neutrophil count ✓ Nausea ✓ Vomiting ✓ Abdominal pain
5.	Ritonavir	155213-67-5	720.948	$C_{37}H_{48}N_6O_5S_2$		✓ drowsiness. ✓ diarrhea. ✓ gas. ✓ heartburn. ✓ change in ability to taste food. ✓ headache. ✓ numbness, burning, or tingling of the hands, feet, or area around the mouth. ✓ muscle or joint pain.

| 6. | Chloroquine | 54-05-7 | 319.872 | $C_{18}H_{26}ClN_3$ | | ✓ blurred vision
✓ difficulty hearing
✓ ringing in ears
✓ muscle weakness
✓ drowsiness
✓ vomiting
✓ irregular heartbeats
✓ convulsions
✓ difficulty breathing
✓ mood or mental changes
✓ decreased consciousness or loss of counsciousness
✓ thinking about harming or killing yourself |
| 7. | Azithromycin | 70288-86-7 | 749.0 | $C_{48}H_{74}O_{14}$ | | ✓ Feeling sick (nausea)
✓ Diarrhoea
✓ Being sick (vomiting)
✓ Losing your appetite
✓ Headaches
✓ Feeling dizzy or tired
✓ Changes to your sense of taste |

With the constant finding of medications and vaccines, shut downs proved to be crucial in reducing virus spread. Most countries have implemented lockdowns to safeguard people, reduce virus spread, and guarantee hospital space. Shutdowns interrupted everyday life around the globe, reducing the intensity and regularity of human activity and output. It played a beneficial role in reducing viral transmission and keeping the environment clean and pollution-free, but it also leads to a decline in the country's economy, with a large number of people becoming jobless. With shifting regulations and levels of confinement, more emphasis is placed on screening and social distancing. COVID-19 had a conflicting effect on our lives and nature, as personal health deteriorated day by day. A cleaner environment was noticed as a result of the shutdown, as businesses were shut down and there was no traffic on the roads.[4-6] The following effects on various external variables are discussed:

Impact on Water Quality

Water utilities all over the globe were impacted by the COVID-19 pandemic. The main industrial contributors of contamination have diminished or halted entirely during the lockdown, which has assisted in lowering the pollution load. As a result of the lack of contamination from industry during India's lockdown days, the rivers Ganga and Yamuna, for example, have attained a high degree of purity. Additionally, it was discovered that throughout the lockdown, as compared to before the lockdown, the concentrations of pH, electric conductivity (EC), DO, BOD, and chemical oxygen demand (COD) had decreased by approximately 1-10%, 33-66%, 45-90%, and 33-82%, respectively, in various measuring sites. Additionally, many areas saw a decline in tourism and water-related activities as a result of the prohibition on public gatherings. Large amounts of solid waste are typically produced during the production and building processes, which also decreases the potential for causing soil and water pollution. Additionally, because of the decline in outsource and import activity, fewer commercial ships and other vessels are moving around the world, which lowers emissions and marine pollution.[7-8]

Impact on Air Quality

A sharp decline in emissions of greenhouse gas (GHG) has been caused by the closure of commercial enterprises, transportation, and industry sectors. As a result of the steps made to combat the the disease the pollutants in the air in New York have dropped by almost fifty percent in contrast to this time previous year. According to estimates, the closure of China's massive manufacturing sectors caused a decrease of nearly fifty percent in N_2O and CO emissions. Another important measure of the world's economy is NO_2 emissions, which show a sign of decline in many nations as a result of the recent lockdown. Typically, consuming fossil fuels releases NO_2, eighty percent of which originates from automobile exhaust. According to reports, NO_2 interacts with oxygen and water to produce acid rain, which in turn causes a number of human illnesses related to breathing. The shutdown imposed by COVID-19 was expected to reduce NO_2 emissions by between thirty and sixty percent in many European areas, including Barcelona, Madrid, Milan, Rome, and Paris, according to the European Environmental Agency (EEA). Automobiles and aircraft are thought to be the main sources of carbon dioxide emissions, accounting for roughly 72% and 11% of the GHG emissions in the transportation industry, accordingly. The worldwide efforts to stamp out the influenza virus are additionally having a significant effect on the aerospace sector. Many nations imposed entrance and exit restrictions on foreign visitors. Commercial aircraft firms are canceling international flights as a result of lower numbers of travelers and regulations. In China, for example, the worldwide outbreak lowers departure capacity by roughly between fifty percent and nine and domestic flight capacity by approximately 70 percent, compared with January 20, 2020, which eventually results in a reduction of almost seventeen percent in nationwide carbon dioxide (CO_2) produced. According to reports, India's overall power generation decreased by nineteen percent after the lockdown, and coal- based power generation fell by twenty-six percent Chinese consumption of coal, which is the greatest in the world, decreased 36% from the same period the year before. (early February to mid-march)[9-10]

Inpact on Soil and Ecological System

Global clinical hazardous material creation has tripled ever since COVID-19 crisis, posing a serious hazard to both the environment and human health. Many pathogenic and biological waste products arise from healthcare for specimen collection of suspicious COVID-19 sufferers, detection, treatment of a lot of people, and disinfection purposes. A significant quantity of healthcare waste is produced every day, the majority of which is not compostable, and its disposal has an adverse effect on the soil. Isolation laws implemented in numerous nations as a result of the epidemic have increased the popularity of internet shopping with delivery service, which has increased the quantity of household garbage produced by sent package materials. Owing to this extensive use and interruption of regular waste disposal, waste recovery, and composting processes, there is an increase in global environmental pollutants and landfilling.[11-12]

Impact on Noice Pollution

The term "noise pollution" refers to the loud sounds produced by various human activities (such as using machinery, driving a car, or doing building projects), that could have a negative impact on people as well as other living things. Noise typically has a bad impact on physical wellbeing, coupled with heart disease, high blood pressure, and inadequate sleep in people. A second way that human excessive noise harms biodiversity would be through changing the equilibrium between prey and predator detection and avoidance. Level of noise has a negative impact on invertebrates as well, which are important for maintaining the ecosphere balance via regulating environmental processes. The isolation and lockdown procedures, on the other hand, forced people to stay inside and decreased worldwide economic activity and communication, which in turn will reduce noise levels in most cities. The amount of noise at Govindpuri metro station (Delhi) have been decreased about 100 dB to 50-60 dB due to diminished car activity throughout the shutdown time. Furthermore, due to constraints on travel, the frequency of flights and automobiles on the roads has decreased dramatically around the world, lowering the overall level of noise pollution. In Germany, for example, passenger flight journey was reduced by over ninety percent, automobile traffic has endured reduced by over fifty percent, and railroads are operating at twenty-five percent less than normal rates. As a whole, the COVID-19 shutdown and decreased commercial activity cut noise pollution around the world. The Govindpuri metro station in Delhi has 50–60 dB less noise than it had before the shutdown because of a drop in vehicle mobility. Also, fewer airlines and transportation vehicles globally as a result of immigration limitations have significantly decreased the amount of noise pollution. For instance, passenger air travel has decreased by over 90percent in Germany, while car traffic has down by over 50percent and train fares are operating about 25percent less than normal. In general, the COVID-19 lockout and a decline in economic activity lowered traffic noise worldwide.[13-14]

Conclusion

This research provided a comprehensive evaluation of the latest research on the natural causes and effects of COVID-19. We investigated the problem from two perspectives: the effect of

COVID-19 on our surroundings and the consequence of environmental markers on the spread of COVID-19 and fatalities. The latest research's results corroborate that the worldwide outbreak of COVID-19 has caused serious socioeconomic and social disruption in all established as well as emerging nations. Despite the fact that they took several precautions, the proliferation of COVID-19 is inevitable. Authorities across the globe are making efforts to avoid virus transmission by means of quarantine, isolation from society, and psyche-hygiene practises. All of these inappropriate disposals have posed a severe danger for waste control. As a result, the governing body ought to create strategies to deal with infectious diseases, and garbage management decisions need to be revised to accommodate rising waste volumes.

References

[1] Facciolà, A., Laganà, P., & Caruso, G. (2021). The COVID-19 pandemic and its implications on the environment. *Environmental research*, *201*, 111648.

[2] Cheval, S., Mihai Adamescu, C., Georgiadis, T., Herrnegger, M., Piticar, A., & Legates, D. R. (2020). Observed and potential impacts of the COVID-19 pandemic on the environment. *International journal of environmental research and public health*, *17*(11), 4140.

[3] Rume, T., & Islam, S. D. U. (2020). Environmental effects of COVID-19 pandemic and potential strategies of sustainability. *Heliyon*, *6*(9), e04965.

[4] Muhammad, S., Long, X., & Salman, M. (2020). COVID-19 pandemic and environmental pollution: A blessing in disguise?. *Science of the total environment*, *728*, 138820.

[5] Saadat, S., Rawtani, D., & Hussain, C. M. (2020). Environmental perspective of COVID-19. *Science of the Total environment*, *728*, 138870.

[6] Sarkis, J. (2020). Supply chain sustainability: learning from the COVID-19 pandemic. *International Journal of Operations & Production Management*, *41*(1), 63-73.

[7] Feizizadeh, B., Omarzadeh, D., Ronagh, Z., Sharifi, A.,

Blaschke, T., & Lakes, T. (2021). A scenario-based approach for urban water management in the context of the COVID-19 pandemic and a case study for the Tabriz metropolitan area, Iran. *Science of The Total Environment*, *790*, 148272.

[8] Sivakumar, B. (2021). COVID-19 and water. *Stochastic Environmental Research and Risk Assessment*, *35*(3), 531-534.

[9] Rita, E., Chizoo, E., & Cyril, U. S. (2021). Sustaining COVID-19 pandemic lockdown era air pollution impact through utilization of more renewable energy resources. *Heliyon*, *7*(7), e07455.

[10] Suthar, S., Das, S., Nagpure, A., Madhurantakam, C., Tiwari, S. B., Gahlot, P., & Tyagi, V. K. (2021). Epidemiology and diagnosis, environmental resources quality and socio- economic perspectives for COVID-19 pandemic. *Journal of environmental management*, *280*, 111700.

[11] Lal, R., Brevik, E. C., Dawson, L., Field, D., Glaser, B., Hartemink, A. E., ... & Sánchez, L. B. R. (2020). Managing soils for recovering from the COVID-19 pandemic. *Soil Systems*, *4*(3), 46.

[12] Singh, A. P. (2021). Covid-19 pandemic its impact on earth, economy and environment. *Asian Journal of Current Research*, *6*(2), 37-41.

[13] Puglisi, G. E., Di Blasio, S., Shtrepi, L., & Astolfi, A. (2021). Remote working in the COVID-19 pandemic: Results from a questionnaire on the perceived noise annoyance. *Frontiers in Built Environment*, *7*, 688484.

[14] Syaiful, S., Siregar, H., Rustiadi, E., Hariyadi, E. S., Prayudyanto, M. N., & Rulhendri, R. (2022). Noise from the traffic volume of motorcycle during the Covid-19 pandemic: A case study of Wiyata Mandala Junior High Schoool Bogor. *Sustinere: Journal of Environment and Sustainability*, *6*(1), 44-54.

Department of Chemistry,
Banasthali Vidhyapith, Rajasthan, India
*Corresponding author : rajendra1509@gmail.com

4. The Role of Technology in Developing Communication Skills in the Digital Age

Prof. Zarreen Naz Mohammed Iqbal

Abstract

The purpose of this research paper is to explore the role of technology in developing communication skills in the digital age. Specifically, this paper will examine the ways in which technology can be used to enhance communication skills, as well as the potential drawbacks and challenges associated with this approach. The purpose of this study is to identify key findings, trends, and gaps in knowledge related to the impact of technology on communication skills development. The literature review examines existing research on effective communication skills in personal and professional contexts, the potential risks associated with over-reliance on technology, and the ways in which technology can be used to enhance communication skills. Research has found that online communication tools, virtual meeting software, and social media can provide individuals with opportunities to practice their communication skills, collaborate with others, and build relationships. However, there are also potential drawbacks associated with relying too heavily on technology, such as a decrease in face-to-face communication skills and neglecting to develop other important communication skills. By considering the potential benefits and limitations of technology use, individuals can become more effective communicators in both personal and professional contexts in the digital age.

Keywords : *technology, communication skills, digital age, effective, communication, online communication, virtual meeting software, social media, over-reliance*

Introduction

In the digital age, technology has become an essential tool for communication. From social media platforms to virtual meeting software, technology has revolutionized the way we communicate with each other. As a result, it has also become increasingly important to develop effective communication skills that are suited to the digital landscape. One of the primary benefits of using technology to develop communication skills is that it provides individuals with the opportunity to practice and refine their skills in a safe and controlled environment. For example, online communication tools such as **chat rooms** and **discussion forums** allow individuals to engage in conversations with others without the pressure of face-to-face interaction. This can be particularly helpful for individuals who struggle with social anxiety or who may feel intimidated by traditional forms of communication. Another advantage of using technology to develop communication skills is that it provides individuals with access to a wider range of communication opportunities. With the advent of virtual meeting software, individuals can now participate in meetings and discussions from anywhere in the world. This can be particularly beneficial for companies with a global workforce, as it allows employees to collaborate and communicate more effectively across different time zones and geographic locations.

However, there are also potential drawbacks and challenges associated with using technology to develop communication skills. For example, some individuals may rely too heavily on technology to communicate and may struggle with face-to-face interactions as a result. Additionally, there is a risk that individuals may become too reliant on technology and may neglect to develop other important communication skills, such as active listening and nonverbal communication. The role of technology in developing communication skills in the digital age is a complex and multifaceted issue. While there are clear advantages to using technology to enhance communication skills, it is important to recognize the potential drawbacks and challenges associated with this approach. By examining these issues in greater depth, this paper aims to provide a more nuanced understanding of the ways in which technology can be used to develop effective communication skills in the digital age.

Review of Literature :
The review of literature for the research paper, "The Role of Technology in Developing Communication Skills in the Digital

Age" examines existing research on the topic of communication skills development and the impact of technology on these skills. The literature review aims to identify key findings, trends, and gaps in knowledge related to this topic. Several studies have highlighted the importance of effective communication skills in the workplace and in personal relationships. Effective communication has been shown to be a critical factor in building trust, resolving conflicts, and achieving goals. However, there is also evidence to suggest that individuals are becoming increasingly reliant on technology to communicate, which could have negative impacts on their ability to communicate effectively in face-to-face interactions.

A number of studies have explored the relationship between technology use and communication skills development. Research has found that online communication tools, such as discussion forums and chat rooms, can provide individuals with a safe and controlled environment to practice their communication skills. Additionally, virtual meeting software has been shown to facilitate more effective collaboration and communication among geographically dispersed teams. However, there are also potential drawbacks associated with using technology to develop communication skills. For example, some studies have suggested that individuals who rely too heavily on technology to communicate may struggle with face-to-face interactions. Additionally, there is a risk that individuals may become too reliant on technology and may neglect to develop other important communication skills, such as active listening and nonverbal communication.

Overall, the existing literature suggests that technology can be a valuable tool for developing communication skills in the digital age. However, it is important to recognize the potential risks associated with over-reliance on technology and to ensure that individuals are developing a broad range of communication skills. Further research is needed to better understand the impact of technology on communication skills development and to identify best practices for using technology to enhance these skills. In addition to the studies mentioned in the literature review, other research has explored the role of technology in developing specific communication skills. For example, several studies have focused on the use of video conferencing technology for developing intercultural communication skills. These studies have found that video conferencing technology can provide opportunities for individuals to communicate with people from different cultures and backgrounds, which can help to develop intercultural competence and improve cross-cultural communication.

Other research has examined the impact of social media on communication skills development. Some studies have suggested that social media use can lead to a decrease in face-to-face communication skills, while others have found that social media can facilitate communication and help to build relationships. While there are clear benefits to using technology to develop communication skills, it is important to note that not all forms of technology may be equally effective. For example, some studies have found that text-based communication (such as email and instant messaging) may be less effective at developing certain communication skills, such as active listening and nonverbal communication, compared to face-to-face interactions.

Overall, the existing research suggests that technology can play an important role in developing communication skills in the digital age. However, it is important to consider the potential drawbacks and limitations of technology and to ensure that individuals are developing a broad range of communication skills. By doing so, individuals can become more effective communicators in both personal and professional contexts, and can navigate the challenges and opportunities of the digital age with greater ease.

Conclusion :

In conclusion, the role of technology in developing communication skills in the digital age is a complex and multifaceted issue. While there are clear benefits to using technology to enhance communication skills, such as providing a safe and controlled environment for practice and facilitating collaboration among geographically dispersed teams, there are also potential drawbacks associated with over-reliance on technology and neglecting to develop other important communication skills Further research is needed to better understand the impact of technology on specific communication skills, such as intercultural communication and nonverbal communication, and to identify best practices for using technology to enhance these skills. Additionally, more research is needed to examine the impact of emerging technologies, such as virtual reality and artificial intelligence, on communication skills development. By considering the potential benefits and drawbacks of technology use and ensuring that individuals are developing a broad range of communication skills, we can become more effective communicators in both personal and professional contexts, and can navigate the challenges and opportunities of the digital age with greater ease.

References :
➤ Anderson, J. A. (2019). Effective communication skills in personal and professional contexts. Journal of Business Communication, 56(1), 23-34.
➤ Bélanger, F., & Watson-Manheim, M. B. (2011). Virtual teams and intercultural communication competence: A conceptual exploration. Journal of Global Information Management, 19(4), 1-16.
➤ Boyd, D. M., & Ellison, N. B. (2007). Social network sites: Definition, history, and scholarship. Journal of Computer-Mediated Communication, 13(1), 210-230.
➤ Cheng, Y., & Wang, Y. (2015). The impacts of social media use on academic performance among university students: A pilot study. Journal of Educational Technology Development and Exchange, 8(1), 1-14.
➤ Chua, R. Y. J., Chen, S. G., & Knezek, G. (2012). Intercultural communication and global virtual teams. In Handbook of Intercultural Communication (pp. 303-318). De Gruyter Mouton.
➤ Duggan, M., & Smith, A. (2016). Social media use in 2016. Pew Research Center. Retrieved from https://www.pewresearch.org/internet/2016/11/11/social-media-update-2016/
➤ Kock, N., & Gemino, A. (2018). The communication advantage of virtual teams: Overcoming the challenges of virtual

collaboration through instant messaging tools. Journal of Business and Technical Communication, 32(3), 339-366.

➤ Lin, M. F. G., Hoffman, E. S., & Borengasser, C. (2013). Is Facebook creating "iDisorders"? The link between clinical symptoms of psychiatric disorders and technology use, attitudes and anxiety. Computers in Human Behavior, 29(3), 1243-1254.

➤ McQuail, D. (2010). McQuail's Mass Communication Theory. Sage Publications.

➤ Ng, K. H., & Ho, S. S. (2014). Effects of virtual team collaboration technologies on communication: A social presence perspective. International Journal of Human-Computer Interaction, 30(2), 122-137.

➤ Riggio, R. E. (2017). Listening and interpersonal skills. Routledge.

➤ Turkle, S. (2011). Alone together: Why we expect more from technology and less from each other. Basic Books.

Research Scholar,
Department of English,
M.G. V's Arts, Commerce and Science College,
Malegaon City- Nashik, MS,
India
email : zarreennaz@gmail.com

5. Synthesis and Study of Photocatalytic Degradation with Microbial Studies of Copper (II) Surfactant

Vandana Sukhadia

Abstract

Photocatalytic degradation has concerned in scientific commune all over the world owing to its several applications in environment, energy, waste water treatment, pollution control, green chemistry, etc. Photocatalytic degradation has been well thought-out to be a disciplined and quick process for degradation for Copper (II) Mustard Urea complex. Copper (II) Mustard Urea complex has been synthesized and measured through FT-IR, NMR, ESR studies. This article recalls and demonstrates the photocatalytic degradation of Copper (II) Mustard Urea complex by heterogeneous photocatalytic process using ZnO as semiconductor. Reaction rate is preferred at the similar time as the photocatalytic activity, which has been governed by a number of factors. Photo-degradation of Copper soap complex varies with light intensity and also affect rate of degradation in different manner. The degradation was conceded out spectrophotometrically in non- aqueous and non polar solvent benzene. Total degradation was calculated and compared with respect to percent degradation.

Keywords : Copper (II) Mustard Urea complex, Zinc oxide as semiconductor, Photocatalytic degradation, Non-aqueous media, light intensity.

Introduction

Photocatalysis is the acceleration of a photoreaction in the subsistence of catalyst. In catalyzed photolysis, light is captivated by an adsorbed subtract. Photocalysis has become one of the most precious approaches to degrade extremely poisonous naturally produce compounds, like cyanotoxins that cannot be isolated through conventional treatment process [1].

Photocatalytic degradation has been measured to be a well-organized and rapid process for degradation of Copper soap derived from edible and non edible oils. Heterogeneous photocatalysis on semiconductor surfaces has concerned a lot of attention due to application like water disinfection, degradation and complete mineralization of organic contaminants in waste water and purification and water splitting for hydrogen production [1-3]. ZnO nano particles are also predictable to be one of the multifunctional inorganic nanoparticles with effective anti bacterial activity. ZnO nano structures exhibits high catalytic efficiency, strong adsorption ability and Au used more and more frequently in the manufacture of sunscreen [4].

In comparable to this initial participation from photo electrochemistry, photocatalysis acquire precious role from other chemical sub disciplines and grow to be a main regulation owing to joint enhancement of scientist arising from different fields: photochemistry, electrochemistry, analytical chemistry, surface science, electronics and catalysis also [5]. Photocatalytic techniques may create to be sooner and more inexpensive than the conventional techniques of treating pollutant.

Current study involves the degradation of complex derived from Copper (II) Mustard soap with Urea ligand. From the analytical data the stoichiometry of complex has been noticed to be 1:1 (metal: ligand). Magnetic moment studies propose the dimeric nature of complex.[6]

Shape and Structure of Micelles :

A micelle is an aggregate of surfactant molecules dispersed in a liquid colloid. The aggregation method depends, on the amphiphilic species and the state of the system in they are dissolved. Hartley proposed that micelles are spherical with the charge groups situated at the miceller surface [6], whereas, Mc Bain recommended that lamellar and spherical forms coexist [7]. X-ray studies by Harkins *et al* [8].Then suggested the sandwich or lamellar model. Later, Debye and Anacker proposed that micelles are rod shaped rather than spherical or disk like [9].

As shown in Figure-1 Micelles of ionic surfactants are aggregates composed of a compressive core surrounded by a less compressive surface structure [10] and with a rather fluid environment (of viscosity 8-17 centipoise (cP) for solubilized nitrobenzene in SDS and cetyltrimethylammonium bromide micelles) [11]. Copper ion attached to micelles have essentially the same hydration shell near the micellar surface as in bulk phase and do not penetrate into the non polar part of the micelles [12] so the volume change caused by the binding of divalent metal ions to micelles is very small [13].

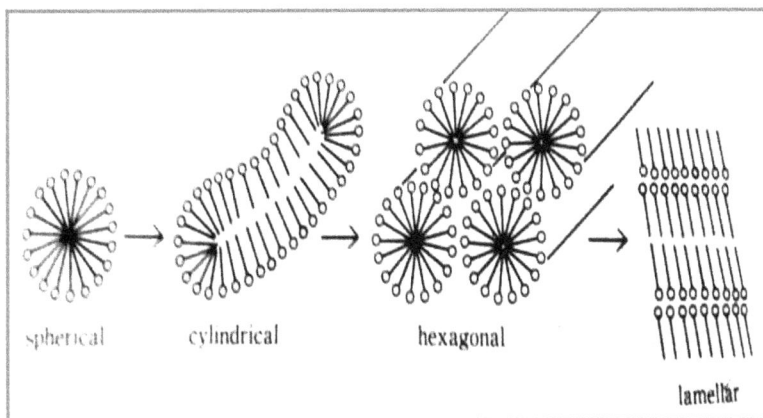

spherical   cylindrical   hexagonal   lamellar

Figure-1 Change in micelle shape and structure with changing surfactant concentration.

Materials and Methods

Initially Copper(II) Mustard soap is equipped by direct metathesis of corresponding potassium hydroxide with oils to get soap with minor excess of requisite amount of Copper sulphate at 50-55°C [14]. Once washing with hot distilled water and alcohol, the sample was dried at 60-80°C and recrystallized with hot benzene.

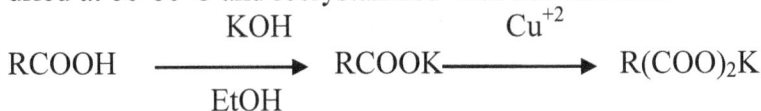

$$RCOOH \xrightarrow[EtOH]{KOH} RCOOK \xrightarrow{Cu^{+2}} R(COO)_2K$$

The synthesized soap derived from edible oil was refluxed with ligand Urea in 1:1 ratio using benzene as solvent for one hour. It was subsequently filtered hot, dried, recrystallized and purified in hot benzene. The complex is dark green and soluble in benzene. Physical parameters are recorded on the basis of its elemental analysis, 1:1 (metal:ligand) type of stoichiometry has been suggested.

Table 1. Fatty acid composition of oil used for Copper soap/complex synthesis [15]

Name of oil	% Fatty acids					
Composition	16:0	18:0	18:1	18:2	18:3	Other acids
Mustard oil	2	1	25	18	10	$(C_{20}$-$C_{41}\%)$

Table 2. The composition and physical data of complex

Name of Complex	Colour	M.P. C°	Yeild %	Metal Content Obs.	Cal.	S.V.	S.E.	Average Molecular Weight
Copper Mustard Urea CMU	Dark green	68	92	8.46	8.35	-	-	759.724

CMU- Copper Mustard Urea complex,
S.V-Saponification value,
S.E- Saponification equivalent

Different amount of catalyst were taken varied from .01, .02, .03, .04, .05 & .06 gm to study the effect of these on the degradation at the same complex solution.

Photocatalytic degradation of CMU complex was recorded at lambda maximum 680nm. Irradiation was carried out in covered glass bottles for the protection of evaporation of solvent with a 200 W tungsten lamp. A water strain was used to keep away from thermal degradation .Concentration of soap complex remains constant during experiment to know the effect of catalyst with the help of solar meter (CEL India Model SM 201). Absorption of light is recorded by U.V. visible spectrophotometer (SYSTRONIC MODEL 106) at different intervals of time.

Result and Discussion

Photocatalytic degradation of Copper Mustard Urea complex was recorded at λ_{max} 680 nm .A plot of 2+ logs O.D. (absorbance) versus time was linear and follows pseudo first order kinetics. Rate of the reaction was calculated using the following expression:

K=2.303x slope

Percent- Degradation of Cmu Complex :

Photocatalytic degradation of CMU complex was carried out by using ZnO as semiconductor under light .Complex degradation was initially identified by color change. Initially the color of complex was dark green- blue which was gradually fades to light green after 2 h. Further light green was disappears slowly and solution becomes almost colorless after completion 18 h light exposure.

Percentage of complex degradation was estimated by the following equation [16].

% degradation= A_o-A_t/A_o*100

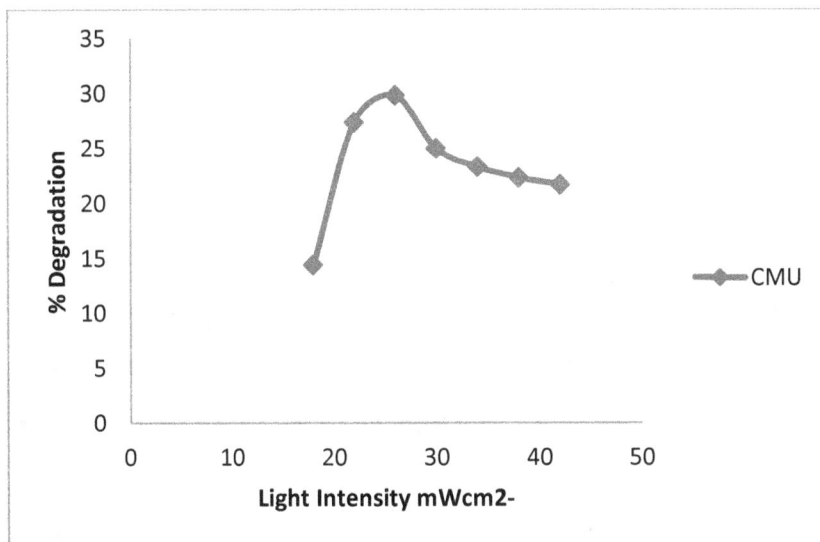

Figure-2 Effect of Light Intensity on Percent Degradation
Effect of light intensity

Photocatalytic degradation of CMU complex was also affected by light intensity. The light intensity was varied from 18, 22, 26, 30, 34 and 42 mW cm^{-2} to 42 mWcm^{-2}. The results are tabulated in Table 3 and presented in Figure-3

The data indicates that the rate of photocatalytic degradation of CMU complex was found to highest at 30 mWcm^{-2}. Further increase in light intensity resulted decrease in the rate of degradation. Value of K1>>K2 shows that the rate of degradation of unsaturated segment is much higher than that of the saturated segment at light intensity 30 mWcm^{-2} as the number of photon striking per unit area of ZnO powder increase with increase in the light intensity, there is a corresponding increase in the rate of photocatalytic degradation of complex . The rate of photocatalytic degradation was found to decrease with further increase in light intensity due to thermal effect. [17, 18, 19]

Solvent   -Benzene,
Amount of ZnO   -0.02 gm,
 [CMU complex]   -Concentration of Copper Mustard Urea
                                               Complex,
Concentration of Complex
in gm mole /lit   - 0.0008M.

Table 3 Effect of Light Intensity on Copper Mustard Urea Complex

Light Intensity (mWcm^{-2})	$K_1 \times 10^{-5}$ sec^{-1}	$K_2 \times 10^{-5}$ sec^{-1}	$K_3 \times 10^{-5}$ sec^{-1}
18	10.03	9.06	13.28
22	12.7	10.2	10.7
26	14.4	9.06	5.09
30	21.9	14.4	17.8
34	17.8	13.6	10.2
38	12.5	12.5	12.28
42	17.1	10.5	6.88

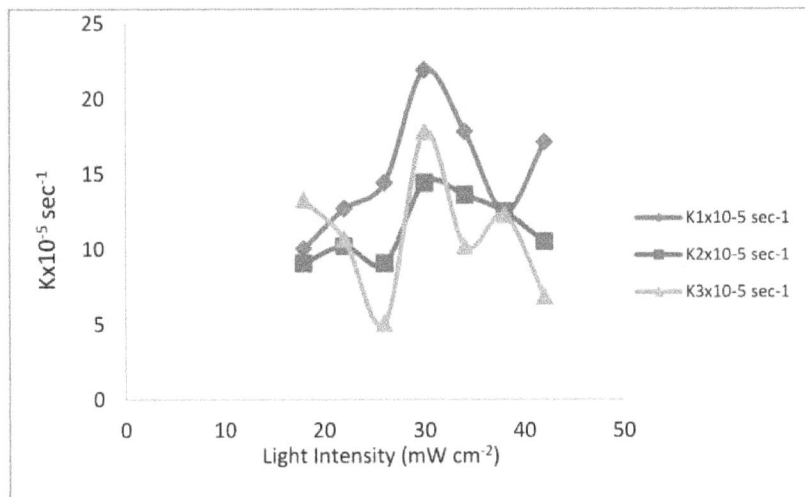

Figure-3 Effect of Light Intensity on Copper Mustard Urea Complex.

Application of the Study

Photodregradation has been considered in a lot of reaction of biological, synthetic and industrial significance here energy received from Sun can be better utilized for converting the pollutants into less toxic or just about harmless materials. Photo degradation plays a vital role in specific phenomena such as foaming, wetting, detergency, emulsification, fungicides etc. The study will endow with much information towards green , safe and sound chemistry.

Conclusion

Current study suggests that the rate of photocatalytic degradation of Copper soap complexes increases by increasing quantity of catalyst at a definite limit and then decreases which may be due to increase in aggregation of macromolecules in solution which decreases the probability and the rate of degradation including various steps of metal ligand breaking, unsaturated segment and saturated segment bond breaking of complex respectively. Rate of degradation increases with increasing amount of semiconductor at some extent and further decreases.

References

1. Hoffemann, M. R., Martin, S. T., Choi, W. Y and Bahemam, D. W., (1995). Environmental Applications of Semiconductor Photocatalysis, *Chem. Rev.*, 95, 69-96. doi: 10.1021/cr00033a004.
2. Mills, A. and Le Hunte, S., (1997) .An overview of semiconductor photocatalysis. *J. Photochem Photobiol A: Chem,* 108, 1-98.
 http://dx.doi.org/10.1016/S1010-6030 (97)00118-4
3. Tryk, D. A., Fujishima, A. and Honda, K. (2000). Recent topics in photoelectrochemistry: achievements and future prospects Author, *Electrochem,   Acta,   45,   2363-2376.* http://dx.doi.org/10.1016/S0013-4686(00)00337-6
4. Ankamwar, B., Choudhary, M., (2005). Gold nano triangles biologically synthesized using Tamrind leaf extract and potential application in vapour sensing, *Synthesis and reactivity in Inorganic and Metal-Organic Chemistry,* 35, 19-26.
5. Herrmann, J. M., (2012). Titania-based true heterogeneous photocatalysis, *Environ Sci Pollut Res,* 3655-3665.
 6. Hartley, G. S., (1936). *Aqueous solution of Parrafin Chain Salts,* Hermann, Paris.
7. McBain, J. W., (1944). Colloid Chemistry, Theoritical and applied J.Alexander,ed), Reinhold, New York.
8. Bloor, D. M., Gormally, J. and Wyn- Jones, E. (1915). *J.Chem. Soc., Faraday Trans.,* 180.
9. Baumuller, W., Hoffman Ulbricht, H. W., Tondre, C. ,and Zana, R. (1978). Solution chemistry of   surfactant, *J. Colloid Interface Sci .,* 64, 418.
10. Mehta, V. P., Talesara, P. R., Sharma, R., Gangwal, A.,and Bhutra, R. (2002).Surface tension studies of ternary system: Copper soap plus benzene plus methanol at 313K, *Indian Journal of Chem. Sec. A,* 41, A(6), 1173-1176.
11. Sharma, R., Tak, P., Saxena, M. , Bhutra, R.and Ojha, K. G.(2008). Synthesis and Characterization of Antifungal Agents Containing Copper(II) Soaps and Derived from Mustard and Soyabean Oil, *Tenside surf. Det.,*45 (2)87-92.
doi.org/10.3139/113.100366

12. Sharma, A., Sharma,V., Mathur, R. P.and Ameta, S. C.(2001). Degradation kinetics of Copper(II) soap derived from pongamia pinnata in presence of irradiating semiconductor ZnO, *Poll. Res.,* 20(3), 419-423.

13. Mattoon, R. H., Stearn, R. S., Harkins, W. D., (1974). Fluorinated Surfactants and Repellents, *J.Phys. Chem.,* 15, 209.

14. Debye P., Anacker, E.W., (1951). Micelle Shape from Dissymmetry Measurements. *J. Phys. Colloid chem.,* 55, 644.

15. Stark, R. V., Kasakevich, M. L., Granger, J. W ,(1982). *Molecularmotio of micellar*
solute;A carbon 13 NMR relaxation study, J. Phys Chem., 86, 335.

16. Vanaja, M., Paulkumar, K., Baburaja, M., Rajesh Kumar, S., Gnanajobitha, G. , Malarkodi, C., Sivakavinesan, M., Annadurai, G.,(2015). *Bioinorganic chemistry and applications.*

17. Sachdeva, D., Parashar, B. , Bhardwaj, S., Punjabi, P. B. , Sharma, V. K., (2010). Use
of *Pure and N, S-Codoped Bimetallic* Cerium Cadmium Oxide Nanoparticles as
Photocatalyst for the Photodegradation of Fast Green, *Int. J. Chem. Sci.,* 8(2), 1321-1328.

18. Ameta, R. , Vardia, J., Punjabi, P. B., Ameta, S. C.,(2006). *Indian J. Chem. Tech.,*
114-116.

19. Upadhyay,R., Sharma, O. P, Jaker, S., Sharma, R. K., Sharma, M. K., (2013).
Photocatalytic degradation of Azure B using Copper hexacyanoferrate as semiconductor,
Int. J. Chem. Sci., 331-340.

Assistant Professor,
Department of Chemistry,
S.D. Govt. College, Beawar (Raj.), INDIA
email : vandana.vandana.sukhadia@gmail.com

6. Sustainable Development and Corporate Sustainability

Dr. Yogita Parihar

Abstract

Sustainable development is a very broad topic in which many other fields like economic growth, environmental protection, social equity, social justice, environmental science and management, business management, policies and law are included. Sustainable development is a process of change in which the exploitation of resources, the direction of investments, the orientation of technological development and institutional change are all in harmony and enhance both current and future potential to meet human needs and aspirations. The objectives of research paper to check real and ground level reality of sustainable development and corporate sustainability. Studies also focus on corporate sustainability work. The study is an exploratory research which is based on secondary sources like books, online research papers and articles, industrial data from different websites, journals, magazines and newspapers etc. For research work, first understand meaning of sustainability than check its implications in business world. The conclusion of research shows that how much sustainability goals are implemented by the industries. The study also check role of business in sustainable development.

Keywords : Sustainability, Corporate Sustainability, Environment, Business, Sustainable Development.

Introduction

Sustainability means an economic equilibrium condition   where the demand place upon the environment by people and commerce can be met without reducing the capacity of the environment to provide for future generations. It indicates development that meets the needs of the present generation without compromising the ability of the future generations to meet their needs. According to World Commission on Environment and Development (WCED) of the UNO – Sustainable development is a process of change in which the exploitation of resources, the direction of investments, the

42

orientation of technological development...instrumental change and the ability of biosphere to absorb the effects of human activities are consistent with future as well as present needs. For sustainable development, government alone cannot play important role. Industry can play very important and key role for wider implication of sustainability. Corporate are driver of economics development so they can play proactive role for social justice and environment protection.

The contribution of Sustainable Development and Corporate Sustainability can understand in two key points –

A) How much industries are affecting environment, social and economic environment.

B) How industries can work with government to achieve predetermine Sustainable Development goals.

Principle of Sustainable Development :

By the world community, there are four fundamental principle given which is written below –

1) Principle of Intergenerational equity – means preserve natural resources for future generation.

2) Principle of sustainable use – means use of natural resources with minimum impact on nature.

3) Principle of equitable use – means equal uses of natural resources by all states.

4) Principle of integration – means social cultural activities and environment are closely integrated.

Corporate Sustainability means industrial development should not cost on natural resources. An industry can maximise its profit but should focus on environmental protection, social justice and equity. It is a business approach that creates long term shareholder value by embracing opportunities and managing risks deriving from economic, environmental and social developments. Corporate sustainability describes business practices built around social and environmental considerations. It is a strategies and practices that aim to meet the needs of the stakeholders today while seeking to protect, support and enhance the human and natural resources that will be needed in the future.

Role of Business in Sustainable Development

Human development is not possible without industries and trading. For Sustainable Development, UN Global Compact is made by United Nations. Global Compact is a strategic policy which is useful for businesses. UN has made ten universally accepted principles for business. These principles are related with human rights, labour, environment and anti-corruption. With the help of ten principles, a business can run its activities for the benefit of economies and societies.

UN made principles are first initiative under which business world in being aligned to common goals such as building markets, combating corruption, safeguarding the environment and ensuring social inclusion and it has resulted in unprecedented partnerships and openness among business, government, civil society, labour and the United Nations. Over 4700 corporate from over 130 countries are participants of Global compact.

Basically Global Compact is a strategically policy which provide framework for development, implementation, and disclosure of sustainability principles. These practices designed to establish sustainable business models and markets building inclusive global economy.

Objectives :

1) In world, total 10 principles are applicable on business activities.
2) Later on, MDGs (Millennium Development Goals) is also included by UN.

Advantages :

- With implementation of these imitative, a common structure for business activities has developed. These policies are related with development, implementation and disclosure of environment, social and governance policies and practices.
- These policy framework solve advance level business problems and helpful to make common strategies.
- It also solves advance level sustainability solution for partnership. Even it solves other than business activities like stakeholder's problems, governments, civil society labour and other non-business activities.

- These principles are useful to create link among businesses and its subsidiaries across the world especially in underdeveloped and growing markets.
- These principles are useful to create extensive knowledge for sustainability.
- Sustainability is the issue which create positive environment for natural resources. It also creates opportunity to do specialized work for environmental, social and governance realms.

UN Global Compact is very much useful and beneficial for private sector. It creates positive opportunities in the environmental, social and governance realms. By partnering with companies in the way and leveraging the expertise and capacities of a range of other stakeholders, the Global Compact seeks to embed markets and societies with universal principles and values for the benefit of all.

Key Drivers regarding Corporate Sustainability

There are four key drivers which guide to ensure sustainability. These key drivers are given below –

1) Internal Capacity Building Strength : It convert various risks into competitive advantage.

2) Social impact assessment : It become sensitive to various social factors like changes in culture, living habits etc.

3) Repositioning capability through development and innovation.

4) Corporate Sustainability : It is a business approach which creates long run shareholders value.

Environmental, economical and social risks can be reduced by above key drivers and create opportunities for businesses. These principles of sustainable development create many opportunities for business units. Some are given below –

- Create good corporate citizen.
- Absolute Value Creation for the Society – when corporate create valuation for the society, they get a opportunity for long run sustainability.
- Ethical Corporate Practices – for long run survival, a corporate should ethically work otherwise it will be shut down soon. In short run, some corporate run their business practices unethically

but it cannot sustain in long run. People deny non-ethical articles which create problem to these corporate.

- A firm can sustain with environmental protection only – natural resources are limited and they have their economical and social value. So these resources should be preserved for long term use.
- Equitable Business Practices – corporate should not involve themselves in unethical activities and should create healthy and fair business environment.
- Corporate Social Responsibility – Corporate is a artificial person. So it is bound to its society wherein they operate and serve. Although there is no hard and fast rules, Corporate Social Responsibility need to be clubbed and integrated into the business model of the company.
- Innovate new technology – innovation is the key to success. Risks and crisis can be eliminated through innovation. Learning and Innovative enterprise gets a cutting edge over others. New innovation is useful to create sustainability but these are useful only that time which innovation is satisfying human need and improve quality of life. New innovations progressively reducing ecological impact and resource intensity to a level at least in line with earth's estimated carrying capacity.
- Market for all – Higher price products, monopoly, unjustified subsidies extra are hindrances to sustainability of a business. Simultaneously, a corporate is to build up its products and services in such a way so as to cater all segments of customers. Customer confidence is essence to corporate success.
- Switching over from stakeholders Dialogue to holistic Partnership – A business enterprises can advance their activities very positively if it makes all stakeholders partner in its progress. It not only builds confidence of various stakeholders, but also helps the management to sheer the business under a very dynamic and flexible system. This approach offers business, government and other stakeholders of the society to build up alliance towards bringing common solutions to common concerns bring faced by all.
- Compliance of statutes – Compliance of statutes, rules and regulations, standards set by various bodies ensure clinical check

up of a corporate and it confers societal license to the corporate to run and operate in the society.

Steps taken by Indian Corporate World under Sustainable Development and Corporate Sustainability :

It is a vision of United Nations to make world free from carbon. So making this reality, all corporates are moving to Zero Greenhouse Gas (GHG). Hopefully, world will achieve this target by 2050. World knows that with transforming world, India too transforming quickly. So India can play key role in this. Currently is a USD $2.72 trillion economy and government has set target of the country to become a USD $5 trillion economy in GDP terms by 2024. To achieve this target, Indian Corporate world is ready. But this achievement cannot be cost on environment. Thus all major Indian Corporates world is ready to achieve target with GHG.

Indian corporate world is adopting Science – based Targets (SBTs) for sustainable development. SBT promotes best practices in emission reductions and net – zero targets in line with climate science. Is also includes a team of experts to provide companies with independent assessment and validation of targets. In India, many companies are using this technique. According to a report by CDP India's Annual Report 2019, total 38 companies are using SBTi, which up from 25 companies the previous year. Now companies are disclosing their information on climate – related risks and opportunities. Business world are also disclosing their environmental control related information with their investors and other stakeholders. It is a good starting by business world, but much more steps yet remaining. Corporates must follow government policies strictly and take government in confidence for zero – carbon future. For this, more investment is invited for zero – carbon future. Aditya Birla Groupm Godrej Grooup, ITC Ltd., Mahindra Group, UPL's technological innovations, Tata Group, Reliance Group etc are implementing zero – carbon base techniques. Following works are given, which companies are using in their premises –

Aditya Birla Group is working on **"<2C Future"**. 2C future is a Paris Agreement commits Business's to limiting global warming to under 2 Degree Celsius above pre-industrial levels. According to this agreement, Aditya Birla Group has developed and followed a logical sustainable development framework. A corporate fundamental working style has been altering and adjusts processes to survive in a hotter world and ensure resilience.

Godrej Group is working on **"Green India"**. It is that initiative which is base on environmental base. The entire value chain of Godrej group is using that process which values environmental related activities. By using Green India initiative, group is capable to reduce GHG emission by 51 percent and more than half of energy resources are base on renewable energy resources.

ITC Ltd. is using **"E – Choupal"** supply system. E- Choupal is supply chain system which is used in agri-business. Thus this system gives direct benefit to farmers. By this system, farmers are able to sale their crop directly to companies, thus formers get more profit on their crop. Thus it is a more transparent process and empowering local people.

Mahindra Group is working on **EV (Electronic Vehicle)**. By using EV, Mahindra Group is transforming Indian transport system. This group is manufacturing EVs at large level with a vision that by 2030, India will become highest EV using country in the world.

UPL's is working on **technological innovations**. These innovations are based on India's food supply chain system. By this innovation, company wants to control wasting habits of Indian.

Tata Group is using **Circular Economy Concept**. This concept is based on "Closing the Loop" principles. This principle is for resources efficiency like waste management by recasting design, process and recycling to create abiding value for their businesses. Under the Closing the Loop initiatives, the group is working on sustainable packaging, producing fertilizers through waste and using industrial by-products in road-construction, fly ash etc.

Reliance Industries is working on **plastic waste management**. For this, recently reliance industry has started a project which is dealing with waste material of plastic in road construction. Industry also has own and outsource garbage collection and segregation system. They also are decomposing this waste.

Conclusion :

Yes, above list of companies are working very good for sustainable development. But still Indian companies efforts are not enough. Many more work yet to do. India is transforming and in transforming journey, corporate support a lot to government. Corporates world are working for identifying climate risks and setting SBT, re – thinking energy strategies to natural climate solutions, by their action plan. In India, many corporate projects are working for sustainable development like EV projects, Renewable Energy projects, food and water related projects etc. Indian companies are good in working for sustainability, risk, and environmental related issues. With government, corporate has positive impacts on sustainable development. A Vision 2050 is also made by world countries. This vision is useful as blueprint for sustainable development.

References :
- Indian businesses can lead global action on sustainability - World Business Council for Sustainable Development (WBCSD)
- In Loop With Success | Tata group
- Reliance to supply waste plastic mixture to strengthen Indian roads | Mint (livemint.com)
- Agricultural Solutions For Farmers | Smart Farming Technology | UPL (upl-ltd.com)
- Mahindra : Home (mahindralastmilemobility.com)
- ITC e-Choupal - Rural India's largest Internet-based intervention (itcportal.com)
- File or directory not found. (godrej.com)
- 2-degrees-Futures-Long.pdf (adityabirla.com)
- WEF_IBC_ESG_Metrics_Discussion_Paper.pdf (weforum.org)
- United Nations in India
- The business of sustainability: Opportunities for India (unep.org)
- India - World Business Council for Sustainable Development (WBCSD)

- Sustainable Development Goals | United Nations in India
- Overview India s Most Sustainable Companies - BW Businessworld
- What India can teach the world about sustainability | World Economic Forum (weforum.org)
- Books on sustainable development
- Journels or magazines
- Seminars and conferences and workshops
- Newspapers and e- newspapers
- Government gadgets and notifications
- Research papers and reviews

Assistant Professor
Business Finance and Economics Department,
JNVU
Shri Sumer Mahila Mahavidhyalya, Jodhpur
email : parihar.yogita@gmail.com

7. Organic Farming A Sustainable Development Practice

Garima Singh

Organic farming is an alternative agricultural system which originated early in the 20th century in reaction to rapidly changing farming practices. It relies on fertilizers of organic origin such as compost manure, green manure and bone meal and places emphasis on techniques such as crop rotation and companion planting. Organic farming permits the use of pyrethrin and rotenone which are naturally occurring pesticides. It also allows the use of synthetic substances such as copper sulphate and elemental sulphur. It prohibits the use of genetically modified organisms, nano materials, human sewage sludge, PGR's, hormones, antibiotics used in livestock and animal husbandry. Organic agricultural methods are internationally regulated and legally enforced by many nations based in large part on the standards set by the **International Federation of Organic Agricultural Movements (IFOAM)**, an international umbrella organization for organic farming established in **1972**.

Organic farming system in India is not new and is being followed from ancient time. It is a method of farming system which is primarily aimed at cultivating the land and raising crops in such a way, as to keep the soil alive and in good health by use of organic wastes (crop, animal and farm wastes, aquatic wastes) and other biological materials along with beneficial microbes (biofertilizers) to release nutrients to crops for increased sustainable production in an eco friendly pollution free environment.

Need of Organic Farming :

With the increase in population our compulsion would be not only to stabilise agricultural production but to increase it further in sustainable manner. The scientists have realised that the green revolution with high input use has reached a plateau and is now sustained with diminishing return of following dividends. Thus a natural balance needs to be maintained at all costs for existence of life and property. The obvious choice for that would be more relevant in the present era, when these agrochemicals which are produced from fossil fuels and are not renewable and are diminishing in availability. It may also cost heavily on our foreign exchange in future.

Organic farming methods can increase farm productivity, repair decades of environmental damage and knit small farm families into more sustainable distribution networks leading to improved food security if they organise themselves in production, certification and marketing. During the last few years an increasing number of farmers have shown lack of interest in farming and the people who used to cultivate are migrating to other areas. Organic farming is one way to promote either self sufficiency or food security. Use of massive inputs of chemical fertilizers and toxic pesticides poisons the land and water heavily. The after- effects of this are severe environmental consequences including loss of topsoil, decrease in soil fertility, surface and groundwater contamination and loss of genetic diversity.

Objectives of Adopting Organic Farming :

1. Increase genetic diversity.
2. Promote more usage of natural pesticides.
3. Ensure the right soil cultivation at the right time.
4. Keep and build good soil structure and fertility.
5. Control pests, diseases and weeds.

Types of Organic Farming :

1. **Pure Organic farming** : It involves the use of organic manures and biopesticides with complete avoidance of inorganic chemicals and pesticides.

2. **Integrated Organic farming** : It involves integrated nutrient management and integrated pest management. It is the type of farming in which the development of crops from natural resources having the complete nutritive value and manage to prevent the crop or plants from the pests.

Techniques used in Organic Farming :

Crop Rotation : It is the technique to grow various kinds of crops in the same area according to different seasons in a sequential manner.

Green Manure : It refers to the dying plants that are uprooted and turned into the soil to make them act as a nutrient for the soil to increase its quality.

Biological pest control : Use of living organisms to control pests with or without the use of chemicals.

Compost : Highly rich in nutrients, it is a recycled organic matter used as a fertiliser in the agricultural farms.

Methods of Organic Farming :

1. Soil Management : After cultivation of crops, the soil loses its nutrients and its quality depletes. Here, use of bacteria present in the animal waste replenishes its nutrients making it more productive and fertile.

2. Weed management : Organic farming helps in lowering the weed rather than removing it completely. This is done by two methods.

Mulching- use of plastic films or plant residue on the surface of the soil to block the growth of the weed.

Mowing/ Cutting- removal of weeds top growth.

3. Crop diversity : Polyculture or Crop Rotation.

4. Controlling the other organisms.

Importance of Organic Farming :

1. Environmental benefits- Natural habitat sources are less threatened.

2. Provides healthier food for people.

3. Soil is in better condition.

Advantages of Organic Farming :

1. Helps to prevent environmental degradation.

2. Used to regenerate degraded areas.

3. Optimal condition in soil and good quality of crops.

4. Reduce the need for purchased inputs.

5. Improves the soil chemical properties.

6. Organically grown plants are more resistant to diseases hence few chemical sprays are used.

7. Poison free.

Disadvantages of Organic Farming :

1. roduction costs are high so the farmers need more workers.
2. Food illness
3. Organic food is more expensive.
4. Not enough production for world's population.

Principles of Organic Farming :

1. Principle of Health
2. Principle of Ecological Balance
3. Principle of Fairness
4. Principle of Care

Why is conventional farming unsustainable ?

The conventional farming is unsustainable due to several reasons such as :

1. Loss of soil fertility due to excessive use of fertilizers.
2. Nitrates run off during rain contaminates water resources.
3. Use of poisonous biocide sprays to curb pests and weeds.
4. Soil erosion due to deep plowing and heavy rains.
5. Loss of biodiversity due to monoculture.

References :

N.Ravi Sankar, M.A.Ansari, A.S.Panwar, C.S.Aulakh, S.K.Sharma, M.Suganthy, G.Suja, D.Jaganathan, "Organic Farming Research in India : Potential Technologies and Way Forward", *Indian Journal of Agronomy 66* (5th IAC Special Issue) : S142-S162 (2021).

S.K.Yadav, Subash Babu, M.K.Yadav, Kalyan Singh, G.S.Yadav, Suresh Pal, " A review of Organic Farming for Sustainable Agriculture in Northern India" , *International Journal of Agronomy*", (Article ID -718145) (2013).

B.Suresh Reddy, "Organic Farming : Status,Issues and Prospects- A Review", *Agricultural Economics Research Review*- Vol.23, 343-358 (2010).

K.A.Gopinath, Ch.Srinivasa Rao, A.V. Ramanjaneyulu, M.Jayalakshmi, G.Ravindrachary ,G. Venkatesh, " Organic

Farming Research in India : Present Status and Way Forward", *International Journal of Economic Plants* 3 (3), 098-103 (2016).
Cong Thanch Vguyen, Tran Thi Tuyet Van , " Development of Organic Agriculture in the Mekong Delta -Opportunities and Challenges", *European Journal of Development Studies* - Vol1; Issue 2 (2021).
P.Panneerselvam, Niels Halberg, Mette Varst, John Erk Hermansen, "Indian Farmers Experience with and Perceptions of Organic Farming", *Renewable Agriculture and Food Systems*, 1-13 (2011).

**Department of Botany,
Agrawal P.G.College, Jaipur.
email : garimads@rediffmail.com**

8. Review on Medicinally Important Plant : *Dalbergia Melanoxylon* Guill. & Perr.

Ms. Saylee V. Surve
Dr. Shailaja Nair*

The Fabaceae, also known as the pea family, contains over 670 genera and close to 20,000 species of trees, shrubs, and herbs. Numerous plants in the Fabaceae family have been the subject of pharmacological investigations that have revealed their antibacterial, anti-fungal, anti-hypertensive, anti-oxidant, antiviral, insecticidal, diuretic, hypoglycemic, and cytotoxic properties. The genus *Dalbergia* is placed under the family Fabaceae and subfamily Faboideae, which are present mostly in tropical and subtropical regions. Around the world, several species of *Dalbergia* are used in traditional medicine to cure a wide range of illnesses, including gonorrhoea, dyspepsia, dysentery, diarrhoea, loss of vision, scabies, rheumatism, and ringworm. They have traditionally been used for their larvicidal, anti-inflammatory, anthelmintic, and analgesic properties. *Dalbergia melanoxylon Guill and Perr*, a traditional herbal medicinal plant belonging to the family Fabaceae, has long been used in South Africa to treat diarrhoea, headache, bronchitis and colds, rheumatism, prevent miscarriage, treat gonorrhoea, stomachache, and abdominal pain, clean wounds, and relieve pain. This review aims to evaluate the traditional uses, pharmacological, pharmacognostic, anti-oxidative, antimicrobial, and cytotoxic activities of various parts of *D. melanoxylon Guill and Perr.* Perspectives for possible future investigations on *D. melanoxylon* Guill and Perr. are also discussed.

Keyword : *Dalbergia melanoxylon Guill and Perr*, Traditional uses, Phytochemistry and Pharmacology.

Introduction:

Everyone on earth uses plants in some or other way as a medicine.with the aid of research on these therapeutic plants; it help us to greatly improve our health and well being . We still don't fully understand the benefits of this amazing natural gem due to the lack of research on its therapeutic characteristics and toxicity. (Kavita et al.2022)

Pharmaceutical is one of the important branch of science which is also associated with the medicinal science. Today pharmaceutical chemistry and pharmacognosy play important role in curing disease and its prevention (varanasi subhose et al. 2005).

Plants have different chemical compounds, like secondary metabolites, with different biochemical and bio-activity properties. It shows applications in various industries, such as pharmaceuticals. The interest in using medicinal plants is increasing worldwide due to their safety, efficacy, cultural acceptability, and lesser side effects compared to synthetic drugs. At present, more than 80% of the global population depends on traditional plant-based medications for treating various human health problems. (koly aktar et al.2017)

There are 300 species in the genus *Dalbergia*, of which roughly 25 are present in India. Many *Dalbergia* species are highly valued as medicinal trees that are also valued for their beautiful, fragrant, aromatic oil-rich wood. (mamta et al., Neeru V. et al. 2009)

Dalbergia melanoxylon Guill & Perr (Family Leguminoceae and subfamily Papilionoidae) include semi-arid, sub-humid, and tropical low land environments. Although it is frequently found on dry, rocky terrain, mixed deciduous forests and savannas of the coastal region are where it present most abundantly. (Tack, 1962,Washa B et al.2014, Nahashon musiba et al 2015) .

The *Dalbergia melanoxylon* is a small, heavily branched tree that typically reaches heights of 4.5 to 7.5 meters but can occasionally reach heights of up to 15-20 meters. Bark is papery and light grey to greyish brown in colour. Plants have compound, opposite, alternating, and 8–13 leaflets. The blooms grow in dense clusters and are whitish in colour. The fruit is a pod that is 3–7 cm long and has one to two seeds in it.(Orwa et al.2009).

The medicinal properties of *Dalbergia* species include analgesic, anthelmintic, anti-inflammatory, antimicrobial, antipyretic, anti-spermicidal, anti-ulcerogenic, aphrodisiac, astringent, expectorant, and larvicidal effects. The majority of the species have previously been proven by various pharmacological studies, while some others need scientific research to support their traditional applications. (Shah et al. 2016, Neeru V. et al. 2009)

It was considered critical to research the therapeutic characteristics of Dalbergia melanoxylon due to the numerous medical uses of *Dalbergia* species. Thus, phytochemical and pharmacological activity will be shown in the current investigation review to assess traditional usage.

Taxonomical Classification : (APG III 2009)

Family-Fabaceae , **Subfamily** -Faboideae, **Tribe** -Dalbergieae

Bionomial Name - *Dalbergia Melanoxylon* .

Common Name :

African blackwood, African ebony, African ironwood, zebra wood, African grenadillo, Senegal ebony, Zebra wood (English) grenadilha pau preto (Po) mpingo in kiswahili (Sw) grenadille d'afrique (Fr) (Bryce 1967, Hine D.A. 1993, Francis Kiondo et. al. 2014)

Geographic Distribution :

Senegal east to Eritrea, Ethiopia and Kenya, south to nimibian, bostwana, northern south Africa and swaziland. It has been introduced in India and Australia. (Bryce 1967, Shah et. al. 2009).

Plant Description :

A highly-branched, deciduous, spiky shrub or small tree with a height of 12–20 meters. Bark surface is thin, smooth, and whitish to pale grey or greyish brown in colour. The inner bark is angel-pink, and the crown is irregular. The nodes were crowded with new shoots. Spirally imparipinnately compound leaves with 7–17 leaflets, Leaflets are alternate to opposite-leathery, short-hairy below; stipules are 2–6 mm long. Petiolules are 1-2 mm long. axillary or terminal panicle measuring 2–12 cm in length, bisexually papilionaceous. Flowers 1–5 mm long, with a white corolla that has an obovate standard, clawed wings, and a keel; usually 9 stamens, fused into a tube, but free in the top section; ovary is superior. Fruit or pods are flat, thin, elliptic to oblong, sharply pointed, and contain one to four seeds. The shape of seeds is kidney-shaped. (Bryce 1967, Orwa et. al. 2009, Washa B. et al. 2014).

Traditional Uses :

In Senegal uses the stem and root bark used in traditional medicine it's conjunction with baobab or tamarind fruits helps to cure diarrhea. The smoke from burnt roots inhaling treat colds, bronchitis, and headaches. A root decoction is used in East Africa to stop miscarriages and to treat gonorrhoea, stomachaches, and abdominal pain, as well as as an anthelmintic and aphrodisiac. Patients with rheumatism in Sudan are exposed to the smoke from burned stems. To treat wounds, use a bark decoction or bark powder, and to relieve numbness in the points, use a leaf decoction. Inflammations of the mouth and throat can be treated with leaf sap. Additionally included in mixes used to cure a variety of diseases are bark decoction and leaf sap **(Bryce 1967, cleopatra N. K. 2017,** Hine D.A. 1993, **Orwa et. al. 2009,** Francis Kiondo's 2014) .

To relieve asthma, dried leaf material is smoked as a cure. (Chigora 2007). Infusion of leaves is used for rheumatic and cardiac pain. (Mohammed G.E.K.2015) .

Plant and Plant Parts

Fig 1.whole plant of *D.melanoxylo* **Fig 2.** Flowers of *D.melanoxylon*

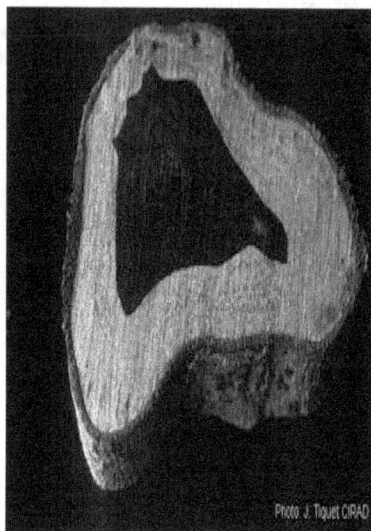

Fig 3. Leaves and pods of *melanoxylon*   **Fig 4.** bark of *D.melanoxylon D.*

Phytochemical Review :

In the heartwood of D. melanoxylon, the quinonoid components (R)- and (S)-4-methoxydalbergione have been identified as the main chemicals. *Dalbergia melanoxylon* bark extracts showed antibacterial and antifungal properties, supporting the plant's conventional use for treating wounds Bryce (1967).

Amri E.et al. (2016) revealed that depending on the type of solvent used, various plant parts show the presence of alkaloids, steroids, flavonoids, saponins, and tannins. Most polar extracts, such as methanol and aqueous extracts, contain more constituents than nonpolar extracts like chloroform and petroleum ether do.

4 dalbergione specifically 3'-hydroxy-4,4'-dimethoxydalbergione, 4-methoxydal- lbergione, and 4'-hydroxy-4-methoxydalbergione, which were separated from ethanol extract of heart wood and then examined for their anti-inflammatory properties .(Feng shao et al. 2022)(Y. misganu et al. 2022).

Meng-fei wang et. al. (2022) isolated 3 new neoflavonoids, entitled (1S,8R,9S),1,5-dihydroxy-4,12-dimethoxy-8-vinyl-tricyclo $7.3.1.0^{2,7}$]trideca-2,4,6,11-tetraen-10- one (1), 2,5,2',5 '-

tetrahydroxy-4-methoxybenzophenone (2) and 2, 5, 3'- trihydroxy-4- methoxybenzophenone (3), from *Dalbergia melanoxylon* heartwood. The broth dilution method was used to test compounds 1–3 for antimicrobial properties against a few microbial strains; however, no chemical showed any significant microbial activity in vitro.

The two new 3-hydroxyisoflavanones (1 and 2) compounds were isolated from the methanol/dichloromethane extract of the stem bark of *Dalbergia melanoxylon* along with 2 known compounds dalbergin and formononetin. It's also shows the presence of flavonoids, terpenes, alkaloids, steroidal saponins, tannins, phenols, and quinines. (Mutai P. et al. 2013, Mohamed G.E.K. 2015).

Qing zhu et. al. (2022) was reported a new chalcone compound, methyl 5-cinnamoyl-2-hydroxy-4-methoxybenzoate (**1**), and a new cinnamylphenol compound, methyl 3-cinnamyl-5-hydroxy-4-methoxybenzoate (**2**) from the CH_2Cl_2 fraction of heart wood *Dalbergia melanoxylon.*

Swetha U. et al. (2017) was detected Total flavonoid content of methanol extract by using Aluminium chloride colorimetric method such as (1mg) equivalent to 5.9µg respectively of quercetin and the total phenolic content of methanol extract(1mg) equivalent to 42.8µg respectively of gallic acid .

Telal M. Najeeb et al. (2018) examined the presence of several phytochemicals in *Dalbergia melanoxylon* leaves and bark extracts made from n-hexane, ethyl acetate, and ethanol. indicated that leaves contained tannins, saponins, sterols/triterpenes, and glycosides but no alkaloids or flavonoids. Additionally, the barks had no saponin but did contain tannins, alkaloids, flavonoids, sterols/triterpenes, and glycosides. Analysis for qualitative detection carried out on extract by using standard procedure as described by **Evans** *et al.*, **Sofowora, Harbone.** The physicochemical properties of leaves and stem were determined according to official methods of analysis of AOAC International (13%) Moisture content crude protein (6.3%), crude fibre (17.4%), and ash (7.4%). Sugar reduction (4.5%) sugars overall (15.2%). The physical characteristics of the bark demonstrate that various extracts, such as n-hexane (0.6%), ethyl acetate (2.6%), and ethanol (5.8%), have varying extractive yields.

Vyas, SV, et al. (2019) studied The physico-chemical parameters of the leaf of Melanoxylon, including the total and acid insoluble ash, are found to be (3.3 wt/%) and (1.47 wt/%), respectively. The foaming index is discovered to be less than 100, while the swelling index is detected as 17.5 units. Also investigated, the phytochemical test showed the presence of different phytochemicals, like tannins, saponins, proteins, reducing sugars, and flavonoids in water extracts, as well as alkaloids, glycosides, and flavonoids in alcohol extracts.

Antimicrobial Activity

Cleopatra N. K. (2017) showed that citric acid extract showed strongest anti-bacterial activity against 7 strains such as *B.subtilis, E.coli, K.pneumoniae, P.aeruginosa, S.typhimurium, S.aureus* and *Y.pestis*. Ethanol extract showed moderate anti-bacterial activity whereas Petroleum ether and Dichloromethane extracts showed no activity. They concluded that bark extracts are potential antiotics.

Amri E. et al. (2016) have revealed that petroleum ether, chloroform, methanol, and aqueous extracts of *D. melanoxylon* root, stem bark, and leaf have antimicrobial activity by using the agar well diffusion method. Stem bark of *D. melanoxylon* showed high inhibitory growth of bacteria and fungi strains compared to other plant parts. Pet ether extract did not show anti- microbial activity. Stem and leaf chloroform and methanol extracts showed higher inhibition against *Candida albicans* while less effective to *S. aureus*. Stem and leaf extracts were more effective rather than root extracts.

Mutai P. et al. (2013) isolated a compound from stem bark using different extracts. The compound was further tested and found effective using the Microtiter-Alamar Blue Assay (MABA) for antimycobacterial activity against *M. tuberculosis* H37Rv (ATCC 27294).

Vasudeva N. et al (2009) has reviewed Dalbergia genus for its antimicrobial activity against gram positive and negative bacteria.

.Maregesi et al. (2008) have claimed, using a liquid dilution method described by Vanden Berghe and Vlietinck, that the n-hexane extract of leaves exhibits mild inhibitory action against Gram-positive and Gram-negative bacteria, such as a *Bacillus cereus, Staphylococcus aureus, Escherichia coli, Pseudomonas aeruginosa,*

Klebsiella pneumoniae, and *Salmonella typhimurium*; as well as antifungal activities on the fungi *Aspergillus niger* and *Candida albicans*.

Telal M. najeeb et al.(2018) examined the results of antimicrobial activity of n-hexane, ethyl acetate and ethanol extracts of *D. melanoxylon* bark by using cup-plate agar diffusion Method. They showed that the highest anti-microbial activity was obtained by Ethanol extract and a moderate activity by ethyl acetate extract against the bacterial strains and *C. albicans*, n-hexane was effective against bacterial strains but showed no activity against *Candida albicans*.

Anti-inflammatory Activity

The isolated compounds like isoflavanones, neoflavone benzofuran and N-cinnamoyl from the acetone extract of heartwood and bark possesses anti-inflammatory activities (Y. misganu et al. 2022).

Qing Zhu et al. (2022) investigated compounds 1 and 2 extracted from heart wood using ethanol against RAW264.7 cells. Compound 1 could considerably decrease LDH activity in 9.6 μM (P< 0.01) and 4.8 μM (P <0.05), whereas the 2.4 and 1.2 μM treatment groups dramatically boosted LDH activity. Compound 2 significantly reduced LDH activity in the 33.7μ M and 16.8 μM groups (P <0.01).

Anti-oxidant Activity

Swetha U. et al. (2017) performed antioxidant activity using a variety of techniques. IC50 values of standard ascorbic acid and *Dalbergia melanoxylon* methanol extract were 19.47 and 30.46 ug/ml for Nitric oxide, 10.22 and 13.35 ug/ml for DPPH, and 27.74 and 33.14 ug/ml for hydrogen peroxide radical scavenging activity, respectively. Hence, the extracts have shown antioxidant activity, according to the above-mentioned observations.

Tejal M. najeeb et al.(2018) revealed that ethanol extract had shown a strong antioxidant activity (76.62%) with IC 50 value of 13.793 μg ml[-1], but n-hexane and ethyl acetate leaves had shown a low anti-oxidant activity with percentages of 11.83% and 20.35%, respectively.

Cytotoxic Activity

Telal M. najeeb et al.(2018) investigated ethyl acetate leaf extract against Culex quinquefasciatus larvae for cytotoxicity activity. It showed significant mortality action at dose concentrations of 1000 ppm and 100 ppm at 308.04 µg ml^{-1} were used to detect the IC 50 value.

Discussion

The current paper discusses scientific research carried out on *D. melanoxylon* with an emphasis on traditional applications, chemical components, and biological activity. The pharmacognostic properties, including physicochemical characteristics and various phytochemical components of the plant, were covered in the current reviews. Also pharmacological effects of *Dalbergia melanoxylon*, included cytotoxic, antibacterial, antioxidant, and anti-inflammatory effects are also discussed. *D. melanoxylon* has been historically used to cure different diseases. Pharmacological research on crude extracts and purified metabolites has realistically supported these traditional applications.

Flavonoids have previously been reported to exhibit a wide range of biological activities like antimicrobial, anti-inflammatory, anti-allergic and antioxidant properties. (Amri E. et al. 2016).

Based on well-established macrophage and fibroblast in vitro bioassays, the three dalbergiones exhibit strong anti-infammatory efficacy. In herbal medicine, these findings pave the way for further research to better understand the underlying molecular mechanism and to translate the findings towards preclinical testing in a relevant model of chronic inflammation such as periodontitis.(Feng shao et al.2022).

The active principles responsible for the therapeutic benefits of medicinal plants are phytochemicals or secondary metabolites of the plant, including but not limited to alkaloids, steroids, flavonoids and tannins .(Telal M. najeeb et al. 2018).

Dalbergia melanoxylon possess a variety of phytochemical constituents that can be isolated for pharmacological activity to improves its medicinal benefits.

Sustainable Consumption and Production

Conclusion

According to the current study, the pharmacological aspects of this species that are studied are antibacterial, antioxidant, anti-inflammatory, and cytotoxic. The future pharmacological aspect can be anti-malarial, anti-diabetic, anti-helminthic, anti-ulcerogenic, etc. This species needs to be the focus of even more research. It is likely that certain known chemicals could end up being the active compounds and prove valuable in the treatment of diseases for which no good cure is known.

Reference :

1) Aktar Koly, Foyzun Tahira(2017) Phytochemistry and Pharmacological Studies of Citrus macroptera : A Medicinal Plant Review,Journal Evidence base complementary and alternative medicine, Volume 2017| Article ID 9789802

2) Amri E, Juma S.(2016) Evaluation of antimicrobial activity and qualitative phytochemical screening of solvent extracts of *Dalbergia melanoxylon* (Guill. and Perr.). Int. Journal of Current Microbiology and Applied Sciences. ;5(7):412-423.

3)*Angiosperm Phylogeny Group (2009)*. "An update of the Angiosperm Phylogeny Group classification for the orders and families of flowering plants: APG III". *Botanical Journal of the Linnean Society. 161 (2): 105–121.* doi:10.1111/j.1095-8339.2009.00996.x.

4) Bryce, J.M. (1967). The commercial timbers of Tanzania. Moshi (Tanzania): Tanzania Forest Division, Utilization Section, p. 139.

5) Chigora, P., Masocha, R., Mutenheri, F. (2007). The role of indigenous medicinal knowledge (IMK) in the treatment of ailments in rural Zimbabwe: the case of Mutirikwi communal lands. J. Sustainable Develop Africa, 9: 26-43.

6) Cleopatra nawa kawanga,(2021). Antimicrobial Activity Of Indigenous Plants Used By Pastoral Communities For Milk Preservation In Kilosa District, Tanzania. *Afribary*. Retrieved from https://afribary.com/works/antimicrobial-activity-of-

indigenous-plants-used-by-pastoral-communities-for-milk-preservation-in-kilosa-district-tanzania.

7) Feng Shao, , Panahipour, L., Omerbasic, A. et al(2022). Dalbergiones lower the inflammatory response in oral cells in vitro. Clin Oral Invest 26, 5419–5428. https://doi.org/10.1007/ s00784-022-04509-7.

8) Francis Kiondo , Tileye Feyissa , Patrick A. Ndakidemi and Miccah Seth,(2014) *In vitro* Propagation of *Dalbergia melanoxylon* Guill. & Perr.: A Multipurpose Tree American Journal of Research Communication : Vol 2 (11), 181-194.

9) Hines D.A., Eckman K, (1993). Indigenous multipurpose Trees of Tanzania: uses and economic benefits to the people. Cultural Survival Canada.

10) Kavita,Nikita Singh and Om Prakash Sharma.(2022), Medicinal plant and their healthy life ,World journal of pharmaceutical and medical research ,*8(9), 197-199.*

11) Mamta Bhattacharya , Archana Singh ,Chhaya Ramrakhyani (2014) *Dalbergia sissoo* - An Important Medical Plant. Journal of Medicinal Plants StudiesYear:, Volume: 2, Issue: 2 First page: (76) Last page: (82) ISSN: 2320-3862 Online Available at www.plantsjournal.com.

12) Meng-fei Wang, Guang-qiang Ma, Feng Shao, Rong-hua Liu, Lan-ying Chen, Yang Liu, Li Yang & Xiao-wei Meng (2022) Neoflavonoids from the heartwood of *Dalbergia melanoxylon*, Natural Product Research, 36:3, 735-741

13) Misganu Y. (2022)Traditional Use, Phytochemistry and Pharmacological Activities of Four Dalbergia Species (Dalbergia Sissoo, Dalbergia Odorifera, Dalbergia melanoxylon and Dalbergia Lactea Vatke): A Review,Volume 23, Issue 7, Page 1-12, Article no.IRJPAC.83311ISSN: 2231-3443, NLM ID: 101647669.

14) Mohamed gamaledin elsadig Karar,(2015). Phytochemical characterization and antimicrobial activity of Sudanese medicinal plants (Doctoral dissertation, Jacobs University Bremen).

15) Mutai P, Heydenreich M, Thoithi G, Mugumbate G, Chibale K, Yenesew A. (2013) Hydroxyisoflavanones from the stem bark

of *Dalbergia melanoxylon*: Isolation, antimycobacterial evaluation and molecular docking studies. Phytochemistry letters.

16) Nahashon Musimba , Josphert N Kimatu, Benard Mweu, MWK Mburu, and Simon Nguluu. (2015). "Germination Effects of Purposive Bruchid Screening of African Ebony (Dalbergia melanoxylon) Seeds in the Arid and Semi-Arid Region of South Eastern Kenya". *Current Research in Agricultural Sciences* 2 (2):60-66.
https://doi.org/10.18488/journal.68/2015.2.2/68.2.60.66.;6(4):6 71-675.*P.*

17) Neeru Vasudeva, Vats, M., Sharma, S.K. and Sardana, S. (2009). Chemistry and biological activities of the genus Dalbergia-A review. Pharmacognosy Rev., 3(6): 307-3014.

18) Orwa C, A Mutua, Kindt R , Jamnadass R, S Anthony. (2009) Agroforestree Database:a tree reference and selection guide version 4.0 (http://www.worldagroforestry.org/ sites/ treedbs/ treedatabases.asp).

19) Qing Zhu , Canyue Ouyang , Yang Liu , Zhangjun Xu , Ying Zhang , Ronghua Liu and Lanying Chen(2022) Anti-inflammatory Continents from the Heartwood of *Dalbergia melanoxylon.Rec. Nat. Prod.* X:X (202X) XX-XX .EISSN:1307-6167

20) Sanjib Saha, Jamil A. Shilpi , Himangsu Mondal , Faroque Hossain, Md. Anisuzzman , Md. Mahadhi Hasan1 , Geoffrey A. Cordell,(2013)Ethnomedicinal, phytochemical, and pharmacological profile of the genus *Dalbergia* L. (Fabaceae)*Phytopharmacology* , 4(2), 291-346.

21) Sheila Mgole Maregesi , Luc Pieters , Olipa David Ngassapaa , Sandra Apers , Rita Vingerhoets , Paul Cos , Dirk A. Vanden Berghec , Arnold J. Vlietinck (2008) ,Screening of some Tanzanian medicinal plants from Bunda district for antibacterial, antifungal and antiviral activities,Journal of Ethnopharmacology 119 58–66.

22) Swetha U.(2017) Antioxidant Activity of *Dalbergia melanoxylon* Bark Extract. International Journal of Applied Pharmaceutical Sciences and Research. ;2*(*4):114- 120.

23) Telal M. Najeeb, Tahani O. Issa, Yahya S. Mohamed, Reem H. Ahmed, Abdelrafie M. Makhawi and Tarig O. Khider,(2018). Phytochemical Screening, Antioxidant and Antimicrobial Activities of Dalberegia melanoxylon Tree,World Applied Sciences Journal36(7):826-833, .ISSN 1818-4952 .

24) Varanasi subhose, pitta srinivas and ala narayan,(2005) Basic principle of pharmacuticle science in ayurveda Bull.Ind.Inst.Hist.Med .Vol XXX-V Page number 83-92.

25) Vyas SV and Naik AA,(2019)Pharmacognostical study on leaf of *Dalbergia melanoxylon* guill and perry .Journal of Pharmacognosy and Phytochemistry ; 8(5): 1404-1407 .E-ISSN: 2278-4136 , P-ISSN: 2349-8234 .

26) Washa, W. B. (2014). A Review of the Literature of Dalbergia melanoxylon. *Int. J. Plant Forest. Sci*, *1*(1), 1-6.Vol. 1, No. 1, , PP: 1 - 6 Available online at http://www.ijp.

27) Yang Liu 2, Ni Zhang , Jun-wei He, Lan-ying Chen , Li Yang1 , Xiao-wei Meng , Feng Shao1 , and Rong-hua Liu,(2021)Two New Compounds From the Heartwood of *Dalbergia melanoxylon* and Their Protective Effect on Hypoxia/Reoxygenation Injury in H9c2.Volume 16(1): 1–7

28) Zhangjun Xu , Yang Liu , Xiaowei Meng , Li Yang , Feng Shao , Ronghua Liu and Lanying Chen(2022) ,Neoflavonoids from the Heartwood of *Dalbergia melanoxylon Record of Natural Product* 16:2 200-205.

Department of Botany,
SVKM'S, Mithibai College of Arts,
Chauhan Institute of Science & Amrutben
Jivanlal College Of Commerce and Economics,
Vile Parle (West), Mumbai, Maharashtra, India.
email : survesaylee21@gmail.com
nairsaila88@gmail.com
*Corresponding Author : nairsaila88@gmail.com

9. पर्यावरण प्रभाव आकलन इकाई की रूपरेखा : महत्व एवं चुनौतियाँ

राजकुमार वर्मा

सारांश :

प्राकृतिक संसाधनों के लगातार ह्रास ने धरातल पर मानव के अस्तित्व के लिए अनेक समस्याएं खड़ी कर दी है। औद्योगीकरण/नगरीकरण में तेजी से वृद्धि और तकनीकी प्रगति ने पर्यावरण का अनेक तरह से शोषण किया है जिसके परिणामस्वरूप गम्भीर ध्वनि, जल व वायु प्रदूषण हुआ है। इसने पर्यावरणीय कानूनों और विनियमों की आवश्यकता का आह्वान किया जो कि पर्यावरणीय प्रभाव आकलन का आधार है जो सतत् वकास के लिए अनिवार्य है। पर्यावरण प्रभाव आकलन (Environmental Impact Assessment) को पर्यावरण पर प्रस्तावित गतिविधि/परियोजना के प्रभाव का पूर्वानुमान करने के लिए अध्ययन के रूप में परिभाषित किया जाता है।

पर्यावरण प्रभाव आंकलन एक निर्णय निर्माण उपकरण है जो एक परियोजना के लिए विभिन्न विकल्पों की तुलना करता है एवं उस एक की पहचान करने का प्रयास करता है जो आर्थिक एवं पर्यावरणीय लागतों तथा लाभों के सर्वोत्तम सयोजन का प्रतिनिधित्व करता है। संक्षेप में हम कह सकते है कि यह निर्णय लेने का एक ऐसा उपकरण होता है जिसमें माध्यम से यह तय किया जा सकता है कि किसी परियोजना को मंजूरी दी जानी चाहिए अथवा नही।

मुख्य शब्द : अस्तित्व, औद्योगीकरण, आकलन, पूर्वानुमान, प्रतिनिधित्व।

पर्यावरणीय प्रभाव आकंलन का इतिहास :

पर्यावरण प्रभाव आकलन की संकल्पना एवं विधि का उद्भव संयुक्त राज्य अमेरिका में 1970 में राष्ट्रीय पर्यावरण नीति अधिनियम पारित होने के साथ हुआ। 1 जनवरी 1970 को राष्ट्रपति रिचर्ड निक्सन ने इस पर हस्ताक्षर किए। राष्ट्रीय पर्यावरण नीति अधिनियम के लागू होते ही भविष्य में क्रियान्वित होने वाली सभी कार्य योजनाओं के लिए पर्यावरणीय प्रभाव आकलन को अनिवार्य बना दिया इसके अभाव में किसी भी प्रस्तावित परियोजना को स्वीकृति नही दी जा सकती थी।

भारत में 1970—80 के दशक में पर्यावरण संरक्षण के लिए कानून बनाने का काम शुरू हुआ। 1972 में स्टॉक होम समझौते में पर्यावरण बचाने के मसौदे पर हस्ताक्षर कर भारत ने इसके प्रति अपनी गम्भीरता जाहिर की। इस समझौते में 1974 में जल और 1981 में वायु से संबंधित कानून बनाए गए। 1984 के भौपाल गैस त्रासदी के बाद एक सयुंक्त पर्यावरण नीति की आवश्यकता महसूस की गई जिसे पर्यावरण संरक्षण अधिनियम 1986 के जरिए पूरा किया गया। इस अधिनियम में वर्णित सन्दर्भों के आलोक में ही 1994 को कुछ विशेष मानदण्ड बनाए गए और बाद में इनमें संशोधन भी किए गए। वर्ष 2006 में इस कानून में कुछ सुधार किए गए। इसके बाद 2020 तक यह कानूनी रूप ले चुका है।

पर्यावरण प्रभाव आकलन की प्रक्रिया : EIA (Enviromental Impact Assessment) प्रक्रिया के 8 चरण हैं –

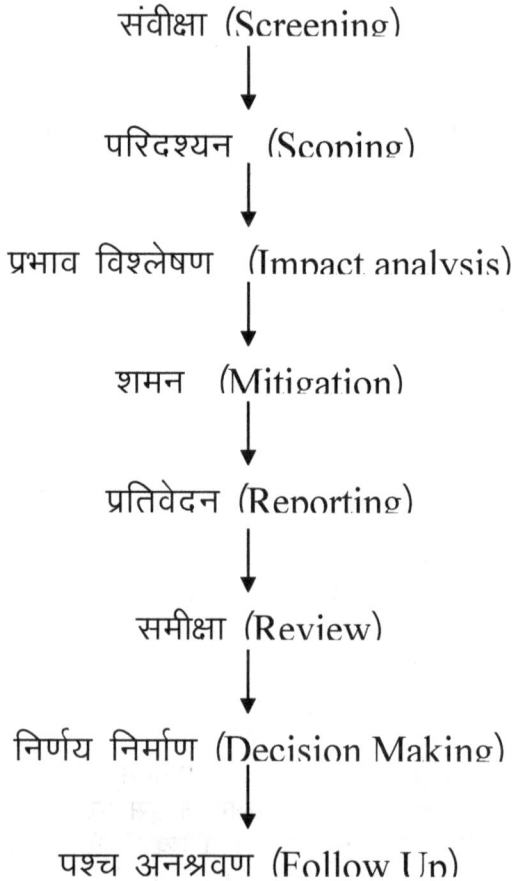

संवीक्षा (Screening)

↓

परिदश्यन (Scoping)

↓

प्रभाव विश्लेषण (Impact analysis)

↓

शमन (Mitigation)

↓

प्रतिवेदन (Reporting)

↓

समीक्षा (Review)

↓

निर्णय निर्माण (Decision Making)

↓

पश्च अनश्रवण (Follow Up)

पर्यावरण प्रभाव आकलन अधिसूचना में 2006 का संशोधन :

1. **परियोजना मंजूरी प्रक्रिया का विकेन्द्रीकरण** : इसके तहत विकासात्मक परियोजनाओं को 2 श्रेणियों में बांटा गया।

- **श्रेणी (।) (राष्ट्रीय स्तरीय मूल्यांकन)** : इसमें परियोजनाओं का मूल्यांकन ''प्रभाव आकलन एजेंसी'' और ''विशेषज्ञ मूल्यांकन समिति'' द्वारा किया जाता है।
- **श्रेणी (ठ) (राज्य स्तरीय मूल्यांकन)** : इसमें परियोजनाओं को ''राज्य स्तरीय पर्यावरण प्रभाव आकलन प्राधिकरण'' द्वारा मंजूरी प्रदान की जाती है।

2. विभिन्न चरणों की शुरूआत : EIA में चार चरणो की शुरूआत की गई। समीक्षा, परिदश्यम, जन सुनवाई और मूल्यांकन।

- श्रेणी (।) परियोजनाओं को अनिवार्य पर्यावरणीय मंजूरी की आवश्यकता होती है। अतः इस तरह विभिन्न संवीक्षा प्रक्रिया से नही गुजरना पडता है।

- श्रेणी (ठ) इसमें परियोजनाएं एक संवीक्षा प्रक्रिया से गुजरती है उन्हे ठ 1 (अनिवार्य रूप से पर्यावरण प्रभाव आकलन की आवश्यकता) और ठ 2 (पर्यावरण प्रभाव आंकलन की आवश्यकता नहीं) के रूप में वर्गीकृत किया जाता है।

3. अनिवार्य मंजूरी वाली परियोजनाएं : खनन, थर्मल पावर, नदी घाटी, बुनियादी अवसंरचना जैसी परियोजनाओं और विभिन्न लघु उद्योगों के लिए पर्यावरण मंजूरी प्राप्त करना अनिवार्य होता है।

पर्यावरण प्रभाव आंकलन अधिसूचना मसौदा 2020 :

पर्यावरण वन एवं जलवायु परिवर्तन मंत्रालय ने पर्यावरण संरक्षण अधिनियम 1986 के तहत मौजूदा पर्यावरण प्रभाव आंकलन अधिसूचना 2006 को प्रतिस्थापित करने के उद्देश्य से पर्यावरण प्रभाव आंकलन अधिसूचना 2020 का मसौदा प्रकाशित किया है जिसमें निम्न प्रस्ताव शामिल है :–

- **जन सुनवाई के लिए आंवटित समय में कटौती** : 2020 में जारी मसौदे में जन सुनवाई के लिये नोटिस की अवधि को 30 दिन से घटाकर 20 दिन करने का प्रस्ताव किया गया है।

- **परियोजनाओं को छूट** : 2020 के मसौदे में "।" "ठ1" और "ठ2" श्रेणी में वर्गीकृत करके इन्हें कुछ छूट प्रदान की गई है।

- **मंजूरी के बाद अनुपालन** : इसका तात्पर्य यह है कि एक बार संबंधित प्राधिकरण द्वारा परियोजना को मंजूरी मिलने के बाद प्रस्तावक परियोजनाओं को EIA रिपोर्ट में निर्धारित कुछ नियमों का पालन करना आवश्यक है ताकि यह सुनिश्चित किया जा सके कि कोई और पर्यावरणीय हानि न हो।

1. 2020 EIA मसौदे में वार्षिक तौर पर अनुपालन रिपोर्ट प्रस्तुत करने का प्रस्ताव है जबकि वर्ष 2006 की अधिसूचना के अनुसार रिपोर्ट हर 6 महीने में प्रस्तुत की जानी थी।

2. अनुपालन रिपोर्ट पूरी तरह से परियोजना प्रस्तावक द्वारा ही तैयार की जायेगी, जो बिना निरीक्षण और समीक्षा के परियोजना में प्रस्तुत गलत जानकारी का कारण बन सकता है।

3. EIA अधिसूचना 2020 में जनता द्वारा उल्लघंन और गैर अनुपालन की रिपोर्टिंग शामिल नही है। इसके बजाय सरकार केवल उल्लंघनकर्त्ता–प्रमोटर, सरकारी प्राधिकरण, मूल्यांकन समिति या नियामक प्राधिकरण से रिपोर्ट का संज्ञान लेगी।

4. EIA 2020 के मसौदे में एक अन्य प्रमुख प्रस्ताव कार्योत्तर मंजूरी देना है, जहां एक परियोजना जो पर्यावरण मंजूरी मिलने से पहले ही कार्य कर रही है

71

को नियमित किया जा सकता है या मंजूरी के लिए आवेदन की अनुमति दी जा सकती है।

5. जिन फर्मो को अपनी स्थापना की शर्तो का उल्लघंन करते हुए पाया गया और अगर उन्हें मंजूरी लेनी है तो जुर्माना देना होगा।

पर्यावरण प्रभाव आकलन का महत्व :

1. यह पर्यावरण संरक्षण और विकास के बीच एक संबंध स्थापित करता है ताकि दोनों में से किसी एक को एक साथ संबोधित किया जा सके और साथ में दीर्घकालिक विकास के उद्देश्य को पूरा किया जा सके।

2. पर्यावरण प्रभाव आकलन (EIA) कम पर्यावरणीय क्षति के साथ परियोजना के लिए एक वैद्य रूपरेखा प्रदान करता है।

3. यह पर्यावरण पर परियोजनाओं के प्रभाव को कम करने के लिए किफायती तरीके प्रदान करा है।

4. पर्यावरण प्रभाव आकलन न्यायपालिका के बोझ को कम करने में मदद करता है क्योकि उचित अध्ययन के अभाव में कई परियोजनाएं कानूनी विवादों का सामना कर सकती है।

5. पर्यावरण प्रभाव आकलन बाढ और भूस्खलन जैसी प्राकृतिक आपदाओं की भविष्यवाणी करने और उनसे बचने मे भी मदद कर सकता है।

6. पर्यावरण प्रभाव आकलन राज्य (सरकार), विकासकत्ताओं, पर्यावरणविदों और जनता के बीच स्वस्थ संबंध बनाए रखने में भी मदद करता है। राज्य और जनता के मध्य अच्छे संबंध बनाए रखना राज्य की शांति और विकास के लिए बहुत महत्वपूर्ण है।

7. पर्यावरण प्रभाव आकलन विकास और पर्यावरण संरक्षण को जोडने में मदद करता है ताकि राज्य सतत विकास के लक्ष्य को प्राप्त कर सके।

पर्यावरण प्रभाव आंकलन की प्रमुख चुनौतिया :

1. पर्यावरण की सुरक्षा के लिए स्थापित पर्यावरणीय प्रभाव आकलन प्रक्रिया पूर्व में भी कई बार सन्देह के घेरे में रही है उदाहरण पर्यावरण पर परियोजनाओं के संभावित हानिकारक प्रभावों से संबंधित म्. प्रक्रिया का आधार प्रायः कम दक्ष सलाहकार एजेंसियां होती है जो इसका प्रयोग कर भ्रष्टाचार को बढावा देती है।

2. पर्यावरण प्रभाव आकलन प्रक्रिया ने उदारीकरण की भावना को न्यून कर दिया है जिससे लालफीताशाही और नौकरशाही को बढावा मिला है।

3. सवैधानिक लोकतंत्र में सरकार को ऐसे कानूनों पर जनता की राय लेनी होती है जिससे बडी संख्या में लोगों के प्रभावित होने की संभावना होती है और कानून के प्रावधानों में उन्हें भागीदार बनाना होता है परन्तु प्रस्तावित मसौदे में सरकार ने जनता के सुझावों के लिए तय समय सीमा को कम करने का प्रयास किया है जो कि चिन्ता का विषय है।

4. सरकार ने प्रस्तावित मसौदे के जरिये अन्य परियोजनाओं के लिए भी ''रणनीति'' राष्ट्र का प्रयोग किया हैं पर्यावरणीय प्रभाव आकलन मसौदा 2020 के तहत अब ऐसी परियोजनाओं के बारे में कोई भी जानकारी सार्वजनिक नही की जाएगी जो इस श्रेणी में आती है इसका सबसे बडा नुकसान यह है कि अब पर्यावरण को प्रतिकूल रूप से प्रभावित करने वाली विभिन्न परियोजनाओं के लिए मार्ग प्रशस्त हो जायेगा।

5. एक चुनौती यह भी है कि विभिन्न देशों की सीमा से 100 किमी की हवाई दूरी वाले क्षेत्र को ''बॉर्डर क्षेत्र'' के रूप में परिभाषित किया जाता है। इसके कारण उत्तर–पूर्व का अधिकांश क्षेत्र इस परिभाषा के दायरे में आ जाएगा जहां पर देश की सबसे अधिक जैव विविधता पाई जाती है, इसके अन्तर्गत सभी अन्तरदेशीय जलमार्ग परियोजनाओं और राष्ट्रीय राजमार्गों के चौडीकरण को पर्यावरण प्रभाव आंकलन मसौदे के तहत मंजूरी लेने के दायरे से बाहर रखा गया है।

6. सरकार के यह सारे प्रावधान पर्यावरण संरक्षण के लिये बने प्रमुख कानून के साथ ही गम्भीर विरोधाभास की स्थिति उत्पन्न करते है।

निष्कर्ष :

पर्यावरणीय मानदण्डों में परिवर्तन से वातावरण पर प्रतिकूल प्रभाव पड़ सकता है। पर्यावरण प्रभाव आकलन 2006 की अधिसूचना पर्यावरण की रक्षा के लिए पर्याप्त शर्त नहीं थी। नवीनतम मसौदे 2020 में कई प्रावधान व्यवसाय करने के लिए, मानदण्डों को आसान बनाने के पक्ष में पैमाने को झुकाते प्रतीत होते है। विश्व बैंक द्वारा 2019 की ''९ विकवपदह इनेपदमे'' रिपोर्ट में भारत 2014 में 142वीं रैंकिंग से लगातार 2019 में 63वीं रैंकिंग पर पहुंच गया है। हालांकि भारत ने पर्यावरण प्रदर्शन सूचकांक पर लगातार गिरावट दर्ज की है। सरकार ने आश्वासन दिया है कि वह पर्यावरण और विकास संबंधी चिन्ताओं के बीच संतुलन बनाने का प्रयास करेगी जिससे जैव विविधता पर गम्भीर प्रभाव ना पड़े क्योकि मानवीय जीवन के गरिमामयी विकास के लिए स्वच्छ पर्यावरण अति आवश्यक है।

सन्दर्भ ग्रन्थ सूची

1. ''मसौदा पर्यावरण प्रभाव आकलन अधिसूचना –2020'' पर्यावरण वन और जलवायु परिवर्तन मंत्रालय भारत सरकार, 12 मार्च 2020।

2. रिचर्ड के. मॉर्गन, ''पर्यावरण प्रभाव मूल्यांकन : कला की स्थिति'' पर्यावरण प्रभाव आंकलन और परियोजना मूल्यांकन 30 जनवरी 2012।

3. ''पर्यावरण प्रभावों का आकलन : विधान की एक वैश्विक समीक्षा'' संयुक्त राष्ट्र पर्यावरण कार्यक्रम, नैरोबी, 2018।

4. ''ड्राफ्ट EIA इन लाइन विद ग्रीन रूल्स, कोर्ट रूलिंग्स : प्रकाश जावडेकर, पर्यावरण मंत्री'' द इकोनॉमिक टाइम्स, 17 अगस्त 2020।

5. ''पर्यावरण मंजूरी नियमों का भारत का प्रस्तावित ओवरहाल मौजूदा नियमों को कमजोर कर सकता है।'' मोंगावे, मार्च 2020।

6. rishtiias.com ''पर्यावरणीय प्रभाव आंकलन : चुनौतियां और महत्व'', 1 अगस्त 2020।

7. drishtiias.com ''जैव विविधता और पर्यावरण प्रभाव आकलन'' 20 मई 2021।

8. ''पर्यावरण मंजूरी और पोस्ट क्लीयरेन्स मॉनिटरिंग पर भारत के नियंत्रक और महालेखा परीक्षक की रिपोर्ट'' भारत के नियंत्रक और महालेखा परीक्षक, भारत सरकार की रिपोर्ट संख्या 39,2016।

9. ''संदीप मित्तल बनाम पर्यावरण, वन जलवायु परिवर्तन मंत्रालय'' नेशनल ग्रीन ट्रिब्यूनल, भारत सरकार, 2020।

10. ''पर्यावरण प्रदर्शन सूचकांक 2020'' पर्यावरण प्रदर्शन सूचकांक, जून 2020।

11. मंजू मेनन और कांची कोहली, ''पर्यावरण गैर अनुपालन को संबोधित करने के लिए नियामक सुधार'' 7 जून 2019, सेन्टर फॉर पॉलिसी रिसर्च।

12. ''पर्यावरण मंत्रालय म्। अधिसूचना के मसौदे पर 17 लाख टिप्पणियां कहता है,'' हिन्दुस्तान टाइम्स, 11 अगस्त 2020।

13. EIA नियमों में संशोधन करके व्यापार के लिए पर्यावरण उल्लंघन को वैध बनाना कार्यकर्ता'' हिमालय वॉचर, 6 जुलाई 2020।

शोधार्थी, भूगोल विभाग
राजकीय लोहिया महाविद्यालय, चुरू (राज.)
email : theverma111@gmail.com

10. Floristic Diversity of Invasive Weeds in Shirpur Taluka of Dhule District, Maharashtra State, India

Rajni Kant Thakur[1,*] and Kumar Ambrish[2]

Abstract

The present study was aimed to document the floristic diversity of invasive weeds in the Shirpur and its adjacent area (Dhule, Nashik, Maharashtra, India). A total of 111 species of weeds belonging to 81 genera and 35 families was recorded in present study. Out of total recorded families, 31 were dicotyledons and 4 monocotyledons. Fabaceae was the dominant family followed by Convolvulaceae, Euphorbiaceae, Malvaceae, Amaranthaceae, Asteraceae, Poaceae, Commelinaceae, Cucurbitaceae, Solanaceae, Apocynaceae, and Boraginaceae. *Ipomoea* (with 8 species) was the largest genus followed by *Euphorbia* (5 species), *Indigofera, Phyllanthus, Sida* (3 species of each), *Alternanthera, Amaranthus, Boerhavia, Calotropis, Commelina, Cynotis, Leucas, Ludwigia, Oxalis, Physalis, Portulaca* and *Senna* (2 species each).

Keywords : Floristic diversity, Crop associated weeds, Invasive species, Nativity, Life-form.

Introduction

A weed is a plant that grows where it is not desired, vying for fertilizer, light, and other resources with cultivated plants. They have characteristic modifications that help in their perpetuation, multiplication, dissemination, stabilization, and overall adaptation (Vasic et al. 2012). The weeds are common dominant, unwanted, undesirable plant that compete with cultivated crop for water, nutrient and sunlight and another several reasons such as, high growth rate, high reproductive rate and produce harmful or beneficial allelopathical effect of cultivated crops (Qasem and Foy 2001). The view of weeds as invasive plants is increasingly shifting around the world as people begin to recognize their importance in broader habitats.

Invasive weed species have characteristics such as being "pioneer species" in different landscapes, being tolerant of a wide range of soil and weather conditions, being a generalist in distribution, producing copious amounts of seed that disperse easily, having aggressive root systems, having a short generation time, high dispersal rates, long flowering and fruiting periods, having a broad native range, and being abundant in their native range. When soil nutrients are lost due to wind and rain, it is these organisms that rapidly establish themselves as the first generation of tough plants in the natural growth of diverse habitats, reducing erosion by the presence of their roots. From the beginning of cultivation, weeds have been called a farmer's worst enemy. Farmers have been fighting them to save their crops for a long time. Invasive weeds inflict dramatic declines in farm, orchard, and grassland production based on their composition and severity. Invasive weeds are the most limiting factors in crop production (Buhler, 1992). Weed exposure is similar to gradual poisoning or disease, with symptoms appearing later in the crop cycle or after harvest. Not only productivity, but also ecological balance, human and animal wellbeing, architectural appeal, and overall economic aspects are all affected. Invasive weed species (IWS) pose a danger to ecosystems, plant species dispersion, and agricultural production.

Weeds, unlike other plants, may withstand severe edaphic, climatic, and biotic conditions. Invasive weed plant research also teaches us about their value, as some of them have a wide range of ethnobotanic applications and may be utilised to produce new pharmaceutical and food items. In other words, 'a weed could be defined a plant out of place or an unwanted plant, or a plant with a negative value, or plants which compete with man for the soil (Muzik, 1970). Many reports are available on the flora of Maharashtra (Singh and Karthikeyan 2000; Singh et al. 2001; Patil 2003, 2010; Sit *et al.*, 2007). No such report, however, is available on the diversity of weeds of Shirpur, Dhule district in Maharashtra. The primary goal of this research was to document the weed flora existing in the Shirpur as baseline information.

Materials and Methods
Study Area
Dhule is district of North Maharashtra (Khandesh area) situated in the lap of Satpura region. Satpura region is well known for its rich biodiversity. The Arunavati river and Tapi river are the rivers flows around the city and fulfill the needs of peoples of Shirpur. Shirpur (21.3496° N latitude; 74.8797° E longitude; 159 m asl altitude) is 50 km from the Dhule. The main profession of the people of the area is agriculture. The main food crops are wheat, barley, maize, finger millet and paddy while sugarcane, cotton, banana, papaya are common cash crops. Besides, the vegetable crops are also cultivated in this region including cucurbits, lady finger, gourd, capsicum, spinach, colocasia, potato, tomato, sugar beet, bean and brinjal.

Methodology
Intensive field studies were conducted to record the maximum number of weeds species in different habitats, i.e., agricultural lands, wastelands, protected areas, river banks and reserve forests of Shirpur during August, 2019 to December, 2022. Villages/ localities visited during the survey include Tarhadi, Shirpur, Warwade, Amode, Abhanpur, Tarhad, Boradi, Dahiwad, Aner, Anturli, Mukhed, Dabhapada, Vakwad, Thalner, Holnanthe and Ziranipada. Plant specimens were collected during the surveys and processed as per the standard method given by Jain and Rao (1976). Small herbs were collected as whole with intact root, stem, leave, flower, and fruit, whereas larger shrubs were sampled as twigs with leave, flower, and fruit. The collected plant specimens were identified with help of available literature, i.e., Hook.f. 1876; Singh and Karthikeyan (2000), Singh et al. (2001) and Patil (2003, 2010) while current nomenclature of plants was adopted from 'Plants of the World Online' database. The nativity of recorded weed species determined following authenticated literature and Plants of the World Online' (2022).

Results and Discussion

A total of 111 invasive weed species belonging to 81 genera and 35 families were recorded from the Shirpur Taluka of Dhule district (**Table 1**). Among 35 families, 31 belong to dicotyledon and 4 to onocotyledon. *Argemone mexicana* L., *Boerhavia diffusa* L., *Cleome viscose* L., *Croton bonplandianus* Baill., *Datura innoxia* Mill., *Euphorbia heterophylla* L., *Euphorbia serpens* Kunth, *Tribulus terrestris* L., *Turnera ulmifolia* L., *Tridax procumbens* (L.) L., *Amaranthus viridis* L., *Cyperus rotundus* L., *Digitari longiflora* (Retz.) Pers., *Cleome viscose* L., *Ludwigia octovalvis* (Jacq.) P.H. Raven, *Phyllanthus urinaria* L., *Senna obtusifolia* (L.) H.S. Irwin & Barneby, *Sonchus asper* (L.) Hill and *Xanthium strumarium* L. were the common weeds in the study area. Some of the plants photographs are shown in below **photoplate 1**. Weeds like *Parthenium hysterophorus* contain several allele chemicals that inhibit the seed germination and growth of other plants (Kumar and Varshney 2007).

Fabaceae is dominant family, followed by Convolvulaceae, Euphorbiaceae, Malvaceae, Amaranthaceae, Asteraceae, Poaceae, Commelinaceae, Cucurbitaceae, Solanaceae, Apocynaceae, Boraginaceae, Brassicaceae, Lamiaceae, Nyctaginaceae, Phyllanthaceae, Onagraceae, Oxalidaceae, Passifloriaceae, Portulacaceae, Verbenaceae, Acanthaceae, Aizoaceae, Amaryllidaceae, Cleomaceae, Cuscutaceae, Cyperaceae, Linderniaceae, Loganiaceae, Martyniaceae, Papaveraceae, Plantaginaceae, Rubiaceae, Scrophulariaceae, Solanaceae and Zygophyllaceae.

The largest genera was *Ipomoea* represented by 8 species, followed by *Euphorbia* (5 species), *Indigofera*, *Phyllanthus*, *Sida* (3 species each), *Alternanthera*, *Amaranthus*, *Boerhavia*, *Calotropis*, *Commelina*, *Cynotis*, *Leucas*, *Ludwigia*, *Oxalis*, *Physalis*, *Portulaca* and *Senna* (2 species each).

Most of weeds were introduced un-intentionally, some introduced due their food values and ornamental purposes from Africa, America, Asia, Mediterranean, Australia, Egypt, Chad, Arabian Peninsula, West Indies, Peru, etc. (**Table 2**).

Farmers have significant challenges in eliminating and managing invasive weeds in their agricultural systems. To remove weeds from agriculture farms, several chemical, biological, and mechanical approaches are used. Weed control can only be successful if the identification, characterisation, and life cycle of weeds are thoroughly understood. Many invasive plants are nevertheless appreciated by individuals who are unaware of their weedy characteristics. Others are identified as weeds, yet property owners do little to stop them from spreading. Some species do not become invasive until they have been ignored for an extended period of time. Invasive plants aren't all created equal. Identification of invasive weeds at the seedling stage is also critical for the successful implementation of an eradication campaign. Weeds' both detrimental and beneficial characteristics must be addressed in any eradication plan. Leguminous weeds, for example, can improve soil fertility by fixing atmospheric nitrogen with the help of some bacteria present in root nodules of these leguminous plants that is beneficial for the crops. Similarly, we must consider how to make appropriate use of weeds that have been removed for diverse purposes. Weeds having therapeutic qualities might be sold to pharmaceutical firms for further research and development. Farmers' income will be increased, either directly or indirectly.

Acknowledgements

The Director, Botanical Survey of India; Director, Shirpur Education Society and Principal, Amrishbhai R. Patel School are acknowledged for administrative help and facilities. We thank our colleague for their support and local people of the Shirpur area for their hospitality and assistance during the field work.

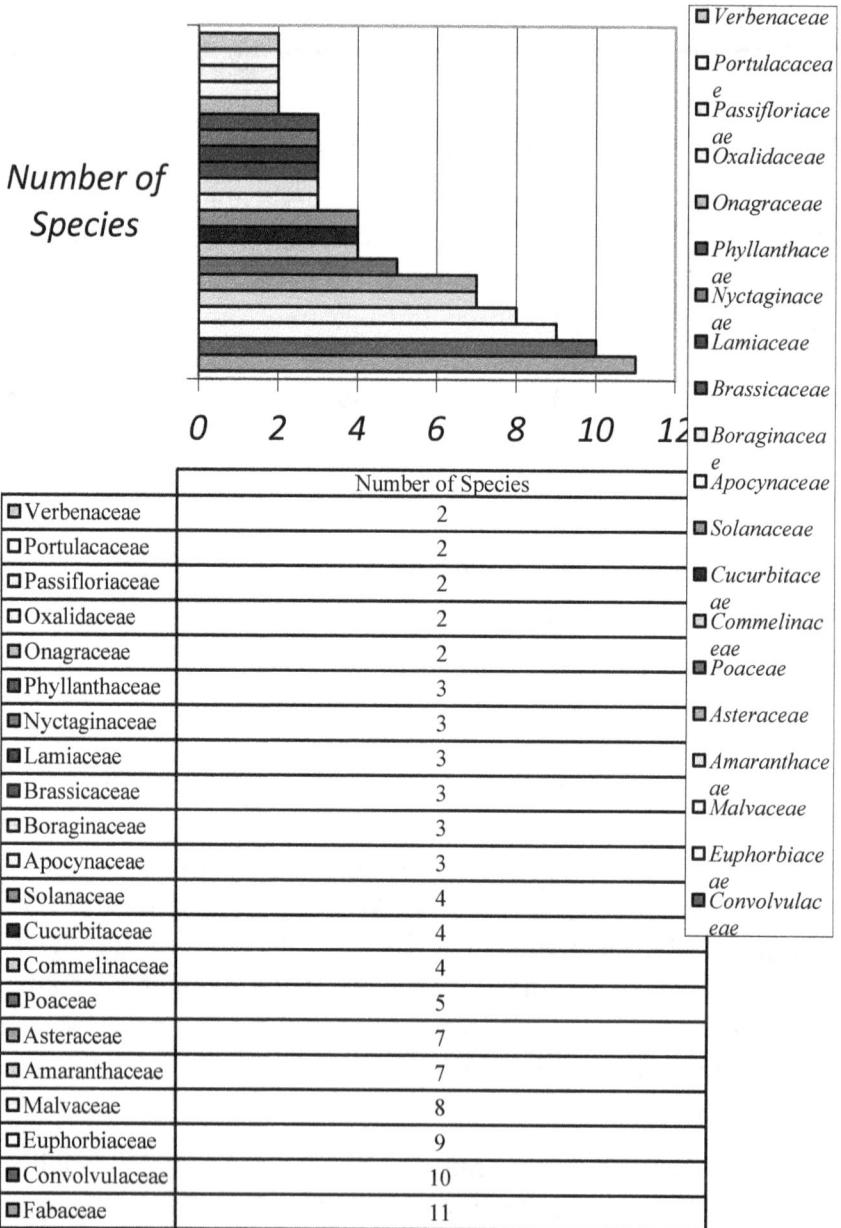

	Number of Species
□ Verbenaceae	2
□ Portulacaceae	2
□ Passifloriaceae	2
□ Oxalidaceae	2
▣ Onagraceae	2
■ Phyllanthaceae	3
■ Nyctaginaceae	3
■ Lamiaceae	3
▣ Brassicaceae	3
□ Boraginaceae	3
□ Apocynaceae	3
▣ Solanaceae	4
■ Cucurbitaceae	4
□ Commelinaceae	4
▣ Poaceae	5
▣ Asteraceae	7
□ Amaranthaceae	7
□ Malvaceae	8
□ Euphorbiaceae	9
▣ Convolvulaceae	10
▣ Fabaceae	11

Fig. 1. Comparative percentage of weed families of Tehsil Shirpur, District Dhule (Maharashtra), India.

References

Buhler CD. 1992. Population dynamics and control of annual weeds in corn as influenced by tillage systems. *Weed Sci.* 40: 241-248.

Hooker , J. D. (1876). *The Flora of British India.* 2: 81-219. L. Reeve & Co., London.

Jain, S.K. & R.R. Rao (1976). *A Handbook of Field and Herbarium Methods.* Today and Tomorrow Printers & Publishers, New Delhi, India, 157pp.

Kumar, S. and Varshney J G., 2007. Biological control of Parthenium : present and future, National Research Centre for Weed Science, Jabalpur, India; pp-157.

Muzik, T.J.(1970). Weed Biology and Control. McGraw Hill Book Co., New York, USA.

Patil DA. 2010. Observations on marsh and aquatic crop Weeds in Khandesh region of Maharashtra. *Life sciences Leaflets.* 2010;10:273-279.

Patil, D.A.(2003). Flora of Dhule and Nandurbar Districts (Maharashtra). Bishen Singh Mahendra Pal Singh, Dehradun, India.

Plants of the World Online (2021) www.plantsoftheworldonline.org

Qasem JR and Foy CL. Weed Allelopathy, its ecological impacts and future prospects : a review. *J Crop Prod.* 2001;4:43-119

Singh N.P. and Karthikeyan S. (2000). Flora of Maharashtra State: Dicotyledons. Volume I. Botanical Survey of India.

Singh N.P., Lakshminarsimhan P., Karthikeyan S. and Prasanna P. V. (2001). Flora of Maharashtra State: Dicotyledons. Volume II. Botanical Survey of India.

Sit AKB. Malay S, Biswanath and Arnunachalam V. 2007. Weed floristic composition in palm gardens in Plains of Eastern Himalayan region of West Bengal. Indian Acad Sci current Sci. 2007;92(10)1434-1439.

Vasic, V., B. Konstantinovic & S. Orlovic (2012). Weeds in forestry and possibilities of their control. In: Price, A.J. (ed.). *Weed Control.*

Table 1: Invasive weeds in Shirpur taluka of Dhule district, Maharashtra state, India

Sr. No.	Name of the Plant	Family	Nativity	Life form	Habit	Mode of Introduction
1	*Abutilon indicum* (L.) Sweet	Malvaceae	Africa	Shrub	Perennial	Ornamental
2	*Acalypha indica* L.	Euphorbiaceae	Tropical & Subtropical Asia.	Herb	A	Un-intentional
3	*Achyranthes aspera* L.	Amaranthaceae	Tropical and sub-tropical Old world	Herb	A	Un-intentional
4	*Alternanthera philoxeroides* (Mart.) Griseb.	Amaranthaceae	Trop. America	Herb	A	Un-intentional
5	*Alternanthera sessilis* (L.) R.Br. ex DC.	Amaranthaceae	Trop. America	Herb	A	Un-intentional
6	*Amaranthus spinosus* L.	Amaranthaceae	Trop. America	Herb	A	Vegetable
7	*Amaranthus viridis* L.	Amaranthaceae	Trop. America	Herb	A	Vegetable
8	*Argemone mexicana* L.	Papaveraceae	S.America	Herb	A	Un-intentional
9	*Boerhavia diffusa* L.	Nyctaginaceae	Tropics & Subtropics.	Herb	Perennial	Medicinal
10	*Boerhavia erecta* L.	Nyctaginaceae	Tropical & Subtropical America.	Herb	Perennial	Un-intentional
11	*Brassica napus* L.	Brassicaceae	S. Europe.	Herb	A	Un-intentional
12	*Buglossoides arvensis* (L.) I.M.Johnst	Boraginaceae	Africa	Herb	A	Un-intentional
13	*Cajanus scarabaeoides* (L.) Thouars	Fabaceae	Asia	Climber	A	Un-intentional
14	*Calotropis gigantea* (L.) Dryand.	Apocynaceae	Trop. America		A	Medicinal
15	*Calotropis procera* (Aiton) Dryand.	Apocynaceae	Trop. America	Shrub	A	Medicinal
16	*Celosia argentea* L.	Amaranthaceae	Trop. Africa	Herb	A	Ornamental
17	*Chrozophora plicata* (Vahl) A.Juss. ex Spreng.	Euphorbiaceae	Africa	Herb	A	Un-intentional
18	*Cleome viscosa* L.	Cleomaceae	Trop. America	Herb	A	Vegetable
19	*Coccinia grandis* (L.) Voigt	Cucurbitaceae	Africa	Climber	A	Vegetable
20	*Commelina benghalensis* L.	Commelinaceae	Tropical & Subtropical Old World	Herb	A	Un-intentional
21	*Commelina forskaolii* Vahl	Commelinaceae	Africa	Herb	A	Un-intentional
22	*Convolvulus arvensis* L.	Convolvulaceae	Temp. & Subtropical Old World	Climber	A	Un-intentional
23	*Corchorus olitorius* L.	Malvaceae	Tropical & Subtropical Old World.	Herb	A	Un-intentional
24	*Crotalaria medicaginea* Lam.	Fabaceae	Asia	Herb	A	Un-intentional
25	*Croton bonplandianus* Baill.	Euphorbiaceae	S.America	Herb	Perennial	Un-intentional

71	*Cucumis maderaspatanus* L.	Cucurbitaceae	Tropical & Subtropical Old World	Climber	A	Un-intentional
26	*Cuscuta reflexa*Roxb.	Cuscutaceae	Mediterranean	Climber	A	Un-intentional
27	*Cyanotis axillaris* (L.) D.Don ex Sweet	Commelinaceae	India to Australia	Herb	A	Un-intentional
28	*Cyanotis cristata* (L.) D.Don	Commelinaceae	NE	Herb	A	Un-intentional
29	*Cynodon dactylon* (L.) Pers.	Poaceae	Trop. America	Herb	Perennial	Un-intentional
30	*Cyperus rotundus* L.	Cyperaceae	Tropical & Subtropical Old World	Herb	A	Un-intentional
31	*Dactyloctenium aegyptium* (L.) Willd.	Poaceae	Tropical & Subtropical Old World.	Herb	A	Un-intentional
32	*Datura innoxia* Mill.	Solanaceae	Trop. America	Shrub	Perennial	Noxious
33	*Descurainia sophia* (L.) Webb ex Prantl	Brassicaceae	Temp. Eurasia	Herb	A	Un-intentional
34	*Digera muricata* (L.) Mart.	Amaranthaceae	SW Asia	Herb	A	Un-intentional
35	*Digitaria longiflora* (Retz.) Pers.	Poaceae	Tropical & Subtropical Old World	Herb	A	Un-intentional
36	*Diplocyclos palmatus* (L.) C.Jeffrey	Cucurbitaceae	Africa and Asia	Climber	A	Un-intentional
37	*Emilia sonchifolia* (L.) DC. ex DC.	Asteraceae	Trop. America	Herb	A	Un-intentional
38	*Euphorbia heterophylla* L.	Euphorbiaceae	Trop. America	Herb	A	Un-intentional
39	*Euphorbia hirta* L.	Euphorbiaceae	Trop. America	Herb	A	Un-intentional
40	*Euphorbia hypericifolia* L.	Euphorbiaceae	Tropical & Subtropical America	Herb	A	Un-intentional
41	*Euphorbia prostrata* Aiton	Euphorbiaceae	Tropical & Subtropical America	Herb	A	Un-intentional
42	*Euphorbia serpens* Kunth	Euphorbiaceae	Tropical & Subtropical America	Herb	A	Un-intentional
43	*Euphorbia thymifolia* L.	Euphorbiaceae	Trop. America	Herb	Perennial	Un-intentional
44	*Heliotropium indicum* L.	Boraginaceae	S.America	Herb	A	Un-intentional
45	*Hyptis suaveolens* (L.) Poit.	Lamiaceae	Trop. America	Shrub	A	Noxious
46	*Indigofera cordifolia* Roth	Fabaceae	Asia	Herb	A	Un-intentional
47	*Indigofera linnaei* Ali	Fabaceae	Trop. America	Shrub	A	Un-intentional
48	*Indigofera trita* L.f.	Fabaceae	Tropical & Subtropical Old World	Climber	A	Un-intentional
49	*Ipomoea triloba* L.	Convolvulaceae	Mexico to Brazil, Caribbean	Climber	A	Un-intentional
50	*Ipomoea cairica* (L.) Sweet	Convolvulaceae	Africa and Asia	Climber	A	Un-intentional
51	*Ipomoea carnea* Jacq.	Convolvulaceae	Trop. America	Shrub	Perennial	Un-intentional

52	*Ipomoea nil* (L.) Roth	Convolvulaceae	Tropical & Subtropical America	Climber	A	Un-intentional
53	*Ipomoea obscura* (L.) Ker-Gaw	Convolvulaceae	Trop. Africa	Climber	Perennial	Un-intentional
54	*Ipomoea pes-tigridis* L.	Convolvulaceae	Trop. East Africa	Climber	A	Un-intentional
55	*Ipomoea purpurea* (L.) Roth	Convolvulaceae	Tropical & Subtropical America	Climber	A	Un-intentional
56	*Ipomoea quamoclit* L.	Convolvulaceae	Trop. America	Climber	Perennial	Un-intentional
57	*Lantana camara* L.	Verbenaceae	Trop. America	Shrub	Perennial	Ornamental
58	*Launaea procumbens* (Roxb.) Ramayya & Rajagopal	Asteraceae	Egypt to Central Asia	Herb	A	Un-intentional
59	*Lepidagathis trinervis* Nees	Acanthaceae	Pakistan to India	Herb	A	Un-intentional
60	*Leucaena leucocephala* (Lam.) de Wit	Fabaceae	Trop. America	Tree	Perennial	Fuel
61	*Leucas aspera* (Willd.) Link	Lamiaceae	Asia	Herb	A	Un-intentional
62	*Leucas longifolia* Benth.	Lamiaceae	W. India, Sri Lanka	Herb	A	Un-intentional
63	*Lindenbergia muraria* (Roxburgh ex D. Don) Brühl	Plantaginaceae	Africa and Asia	Herb	A	Un-intentional
64	*Ludwigia octovalvis* (Jacq.) P.H.Raven	Onagraceae	Trop. America	Herb	A	Un-intentional
65	*Ludwigia perennis* L.	Onagraceae	Trop. America	Herb	A	Un-intentional
66	*Malvastrum coromandelianum* (L.) Garcke	Malvaceae	Trop. America	Herb	A	Un-intentional
67	*Martynia annua* L.	Martyniaceae	Trop. America	Herb	A	Un-intentional
68	*Melilotus officinalis* subsp. *alba* (Medik.) H.Ohashi & Tateishi	Fabaceae	Europe to China, N. Africa to Myanmar, Ethiopia to S. Africa	Herb	A	Un-intentional
69	*Merremia emarginata* (Burm. f.) Hallier f.	Convolvulaceae	Tropical Africa, S. China to Tropical Asia	Herb	A	Un-intentional
70	*Mirabilis jalapa* L.	Nyctaginaceae	Peru	Herb	A	Ornamental
72	*Oxalis corniculata* L.	Oxalidaceae	Europe	Herb	A	Medicinal
73	*Oxalis latifolia* Kunth	Oxalidaceae	Tropical & Subtropical America	Herb	A	Medicinal
74	*Parthenium hysterophorus* L.	Asteraceae	Trop. America	Climber	A	Un-intentional
75	*Passiflora foetida* L.	Passifloraceae	Trop. S. America	Herb	A	Ornamental
76	*Pergularia daemia* (Forssk.) Chiov.	Apocynaceae	Africa and Asia	Climber	A	Ornamental
77	*Phyllanthus maderaspatensis* L.	Phyllanthaceae	Africa and Asia	Herb	A	Un-intentional
78	*Phyllanthus tenellus* Roxb.	Phyllanthaceae	Tanzania to Mozambique, SW. Arabian Peninsula, W. Indian	Herb	A	Un-intentional

			Ocean			
79	*Phyllanthus urinaria* L.	Phyllanthaceae	Tropical & Subtropical Asia to N. Australia.	Herb	A	Medicinal
80	*Physalis angulata* L.	Solanaceae	Trop. America	Herb	A	Un-intentional
81	*Physalis pruinosa* L.	Solanaceae	Trop. America	Herb	A	Un-intentional
82	*Poa annua* L.	Poaceae	Temp. Old World to Tropical	Herb	A	Un-intentional
83	*Portulaca oleracea* L.	Portulacaceae	Trop. S. America	Herb	A	Vegetable
84	*Portulaca quadrifida* L	Portulacaceae	Trop. America	Herb	A	Un-intentional
85	*Prosopis juliflora* (Sw.) DC.	Fabaceae	Mexico	Shrub	Perennial	Un-intentional
86	*Rorippa dubia* (Pers.) H.Hara	Brassicaceae	Indian Subcontinent to China	Herb	A	Un-intentional
87	*Senna alata* (L.) Roxb.	Fabaceae	SW. Mexico to Tropical America	Shrub	Perennial	Ornamental
88	*Senna obtusifolia* (L.) H.S. Irwin &.Barneby	Fabaceae	Tropical & Subtropical America	Shrub	A	Un-intentional
89	*Setaria verticillata* (L.) P.Beauv.	Poaceae	Tropical & Subtropical Old World	Herb	A	Un-intentional
90	*Sida acuta* Burm.f.	Malvaceae	Trop. America	Herb	A	Un-intentional
91	*Sida cordifolia* L.	Malvaceae	Tropical & Subtropical Asia to N. Australia	Herb	A	Un-intentional
92	*Sida rhombifolia* L.	Malvaceae	Tropical & Subtropical Old World	Shrub	A	Un-intentional
93	*Solanum virginianum* L.	Solanaceae	Tropical Africa, Arabian Peninsula, S. Iran to S. Central China and Indo-China.	Herb	A	Un-intentional
94	*Sonchus asper* (L.) Hill	Asteraceae	Mediterranean	Herb	A	Un-intentional
95	*Spermacoce pusilla* Wall.	Rubiaceae	Indian Subcontinent to S. China and Philippines	Herb	A	Un-intentional
96	*Spigelia anthelmia* L.	Loganiaceae	Tropical & Subtropical America	Herb	A	Un-intentional
97	*Stachytarpheta jamaicensis* (L.) Vahl	Verbenaceae	SE. U.S.A. to Tropical America	Herb	A	Un-intentional
98	*Synedrella nodiflora* (L.) Gaertn.	Asteraceae	West Indies	Herb	A	Un-intentional
99	*Tephrosia purpurea* (L.) Pers.	Fabaceae	S. Egypt to Chad, Arabian Peninsula to NW. India	Shrub	A	Un-intentional
100	*Torenia fournieri* Linden ex E. Fourn.	Linderniaceae	India to S. China and Indo-China, Taiwan	Herb	A	Ornamental

101	*Trianthema portulacastrum* L.	Aizoaceae	Tropics & Subtropics	Herb	A	Un-intentional
102	*Tribulus terrestris* L.	Zygophyllaceae	Trop. America	Herb	A	Medicinal
103	*Trichodesma indicum* (L.) Lehm.	Boraginaceae	Trop. Central America	Herb	A	Un-intentional
104	*Trichosanthes cucumerina* L.	Cucurbitaceae	Asia	Climber	A	Un-intentional
105	*Tridax procumbens* (L.) L.	Asteraceae	Trop. America	Herb	A	Un-intentional
106	*Triumfetta rhomboidea* Jacq.	Malvaceae	Trop. America	Herb	A	Un-intentional
107	*Turnera ulmifolia* L.	Passifloriaceae	Asia	Herb	A	Un-intentional
108	*Verbascum coromandelianum* (Vahl) Hub.-Mor.	Scrophulariaceae	Asia	Herb	A	Un-intentional
109	*Waltheria indica* L.	Malvaceae	Trop. America	Herb	A	Un-intentional
110	*Xanthium strumarium* L.	Asteraceae	Trop. America	Shrub	A	Un-intentional
111	*Zephyranthes citrina* Baker	Amaryllidaceae	Central America	Herb	A	Ornamental

[1]Amrishbhai R. Patel School,
Shirpur, Dhule , Maharashtra, India
[2]High Altitude Western Himalayan Regional Centre,
Botanical Survey of India, Solan,
Himachal Pradesh, India
*email corresponding author : thakurrkbsi@gmai.com

11. A Review : Silver Nanoparticles (Ag-NPs) Synthesized from Plant Extracts and Potential Biomedical Applications in Diabetic Foot Ulcer

Dr. Meenu Mangal[*1] and Dr. Sunil Mangal[2]

Abstract

This review includes studies on the silver nanoparticles generated from various plant extracts. The silver ions present in solution of silver nitrate were reduced with the help of reducing agents present in plant extracts. Different components of the plant extracts effectively reduce Ag^+ ions. These may be alkaloids, terpenoids, flavonoids, acids, etc. The different parts of the plants such as leaves, stem, barks, seeds, roots and fruits are used for bringing the Ag+ ions to nano-sized particles. Application of nanoparticles have been done as nanopharmaceuticals, nanodevices, molecular medicines and medical nanorobots. Nanoparticles in Prosthodontics are also used as in acrylic resins, tissue conditioners, dental adhesives, composites, dental cements, implants and maxillo-facial prosthesis. The applications of these Ag-NPs were reviewed for Diabetic foot ulcer, i.e., a non-healing or poorly healing wound below the ankle of the patients with diabetes. The clinical testing of "WinVivo" wound ointment has also been included in this review. The review is of highly importance in the biomedical applications of the silver nanoparticles.

Keywords : Ag-NPs, Diabetic patient, foot ulcer, Antimicrobial, Plant extract, Biomedical

Introduction

Diabetes or hyperglycemia is now being the third top most killer of mankind after cancer and cardiovascular disease. The high prevalence, morbidity and mortality of Diabetes made it upcoming largest threat (WHO, 1999). The diabetic patients in number are rapidly increasing which increases potential burden on the healthcare system [1,2]. Over production of glucose and decreased utilization by the tissues make fundamental basis of hyperglycemic conditions.

The two main type of diabetes is, diabetes mellitus and diabetes insipidus. Marked hyper lipemia that characterizes the diabetes mellitus II state is a result of inhibition of inhibitory action of lipolytic enzymes which acts on the fat deposits. Blood glucose level is controlled by the hormone insulin synthesize by β- cells (islets of langerhans) of pancreas [3]. Diabetes insipidus is caused due to changes in the anti-diuretic hormone secreted by pituitary glands which directly affects water retention in body hence diabetes insipidus is called as "water" diabetes.

On the other hand, diabetes mellitus is often called as "sugar" diabetes and it is caused by the pancreas malfunctioning resulting in insulin deficiency or defect in the secretion of insulin [4]. Diabetes can be broadly classified into three main types: (i) Diabetes mellitus type-I (T1DM) formerly known as insulin-dependent diabetes mellitus is an auto-immune disorder where the body's immune system destroys the β cells in the pancreas. (ii) Diabetes mellitus type-II (T2DM) is the most significant type of diabetes, also known as non-insulin dependent diabetes mellitus.

WHO predicted that Type II diabetes mellitus will soon become a serious health issue worldwide [5]. Due to insulin resistance, glucose does not reach to the cells and hence glucose accumulates in the blood and cause hyperglycemia which is excreted in the urine (glycosuria). (iii) Gestational diabetes (GDM) which is most often seen among pregnant women [6,7].

Diabetic Foot Ulcer

Among the above types of diabetes mellitus, type II DM is most common worldwide. In type II diabetes mellitus patients have abnormalities in sugar level therefore after having meals the sugar level boosts up which is normally termed as post prandial hyper glycaemia.

The continuous swings in glucose level leads to pathological problems and organ failure [4]. The term diabetic foot ulcer refers to "A non-healing or poorly healing full-thickness wound below the ankle with diabetes critical in the natural history of the diabetic foot"[1,7]. Diabetic foot complications, like neuropathy, foot ulceration, peripheral vascular disease and infection with or without osteomyelitis, may lead to serious complications of gangrene and limb amputation.

This serious condition is 25% more common in diabetic patients [2,8]. In a diabetic patient, a diabetic foot ulcer is a wound below the ankle shown in Figure 1. In diabetic patient a foot ulcer causes loss of protective sensation, recurring trauma and ulcers, and poor blood circulation. It may develop further complications, such as multiple bacterial infections [9]. A small infection in acute diabetes can lead to the early development of complications, even in a minor trauma, and this disease develops and becomes resistant to antibacterial therapy.

This infection may arise due to a group of both gram-positive bacteria and gram-negative bacteria [10].The lack of proper sensation predisposes a person with diabetes to suffer from ulcers that do not heal due to poor circulation of blood [11]. Wound healing is further complicated by multiple microbial infections in foot ulcer [12].

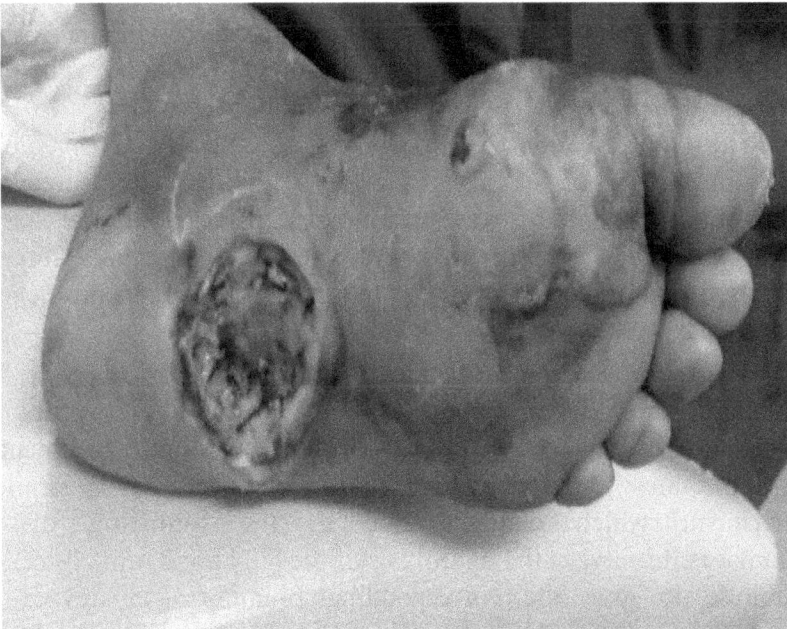

Figure 1. Wound in Diabetic Foot Ulcer patient

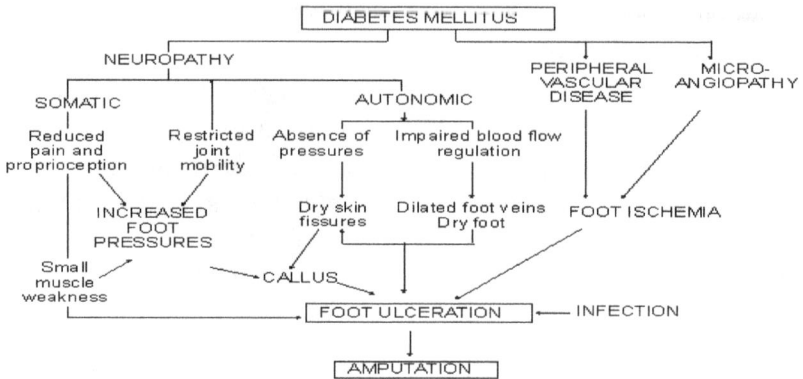

Figure 2. Flow diagram for the effect of Diabetes on foot ulceration

Clinicalaetiology of Diabeticfoot Infections (DFI)

A diabetic foot ulcer involves infections due to aerobes, anaerobes and fungi individually or in combination with other microbes. The infection starts locally with an ulcer affecting immediate surrounding skin with a purulent discharge and erythema. These signs can be missed due to the presence of sensory neuropathy and ischemia which are the two of the commonest risk factors to DFI [13-15]. The infection can become subsequently a spreading infection as the sepsis progresses to cellulitis. This spreading infection can become severe causing extensive deep soft tissue damage. The deep tissue fills with pus and can cause abscess formation subsequently leading to tissue necrosis and severe bacteremia.

The primary goal in the treatment of diabetic foot ulcers is the wound closure. The antibiotic treatment is useful but antibiotic resistance causes problems [16]. Along with antibiotic treatment a multidisciplinary approach needs to be used.There is a familiar mechanism which emerges due to the formation of β-lactamases.Itcatalyzes the hydrolysis of the C-N bond ofthe β-lactam antibiotic to give the corresponding β-amino acid devoid of antibacterial activity [17]. (ii) Bio films are mostly found to be present in two-thirds of chronic wounds and they show resistance to antibiotics [18,19]. Biofilms are defined as large communities of microbial cells growing on a surface and embedded in self-synthesized matrix made up of extra cellular polymeric substances [20]. The major components of these matrixes are the exopolysaccharides with protein, lipids and DNA. Exopolysaccharides have been utilized for the initial attachment and for the development of pathogenic biofilms [20].

Role of Silver

Due to the large use of antibiotics bacteria gained antibacterial resistance over the time. To overcome these limitations clinicians have turned to silver dressings containing varying amounts of silver [21].Silver have antimicrobial effects, including direct inhibition of cellrespiration, intracellular enzymes inactivation and cell membrane alterations [22]. The use of elemental silver for silver-coated dressings may be more effective in killing bacteria than silver compound such as silver nitrate or silver sulfadiazine[23].Silver exhibits its antimicrobial efficiency by interfering with the respiratory chain at the cytochromes [24].Silver ions also interfere with microbial electron transport chain system, bind with DNA and inhibit replication of DNA [25,26]. Silver is effective against fungi, yeast, viruses, a broad range of aerobic, anaerobic, gram positive and gram negative bacteria [23]. Silver also have anti-inflammatory properties suggesting the loss of rubor in chronic wounds when treated with colloidal silver [27-29].

Healing would be promoted by an ideal silver ion donor site dressing material with minimal pain to the patient; prevents infection from pathogenic microbes and result in minimal scarring.A dressing with all of these qualities is desirable but still dressing materials meet some of these criteria to varying degrees.

Silver Nanoparticles

Nanotechnology is the application of nanoscience, engineering and technology to produce nano materials [30].There is an ever growing need to develop clean, non-toxic, and environmentally benign synthesis procedures. Bio-nanotechnology involves production of nanomaterials using plant and animal as well as microorganisms based products.

Researchers are now focusing the synthesis of nanoparticle from biolomaterials. The synthesis of silver nanoparticles from green process is being an important aspect of nanotechnology research. Nanomaterials such as Au, Ag, Pb and Pt have been synthesized by different methods, including hard template, using bacteria, fungi and plants. Among these, silver nanoparticles play an important role in the field of medicine and clinical research due to its physiochemical properties.

Synthesis of Ag-NPs using Plant Extracts

According to a WHO report, 80% of The bulk of traditional medicines use plant extracts or active ingredients, and indigenous medicine is used by the majority of the world's populations [31]. Sliver Nano particles are formed with reduction of silver ions with help of reducing agents. Green synthesisofnanoparticles has numerous advantages in environmental and biomedical fields [32]. This process can be done through a biological or chemical method. Different plant extracts produced the silver nanoparticles (AgNPs) [33]. Green synthesis aims to reduce specifically, the usage of harmful chemicals. Plants also contain reducing and capping agents. Phytochemicals are chemical substances produced by plants as part of their normal metabolic activities. Secondary metabolites include alkaloids, flavonoids, coumarins, glycosides, gums, polysaccharides, phenols, tannins, terpenes, and terpenoids, among other things. The synthesis of silver nanoparticles from different plant extracts have been summarized in Table 1.

Table 1. List of different plant extracts used in synthesis of silver nanoparticles

Name of Plant Extract	Reference	Name of Plant Extract	Reference
Leaf extract of Svensoniahyderobadensis	[34]	Leaf extract of Bixaorellana	[68]
Stem barks of Boswelliashorea	[34]	Leaf extract of Night Jasmine (Nyctanthesarbortristis	[69]
Murrayakoenigii leaf extract	[35]	Leaves and fruit extracts of Securinegaleucopyrus (wild) Mull	[70]
Leaf extractof Vitexnegundo	[36]	Leaf extract of Saracaindica	[71]
Leaf extract of Cleome viscosa	[37]	Leaf extract of Pterocarpussantalinus	[72]

Leaf extract*of Ocimumtenuiflorum*	[38]	*Leaf extract of* Morusnigra	[73]
Leaf extract of Mimosa pudica	[39]	*Leaf extract of Withaniasomnifera*	[74]
Plectranthusamboinicus (Lour.) Spreng aqueous extract	[40]	*Leaf extract of Raphanussativus* var. longipinnatus	[75]
lemon leaf extract (Citrus limon)	[41]	Fruit extract of *Vitisvinifera*	[76]
Leaf extract*of Wrightiatinctoria*	[42]	*Leaf extract of Azadirachtaindica* (Indian Neem).	[77]
Leaf extract of Ananascomosus	[43]	Bamboo leaves extract	[78]
Black pepper corn extract	[44]	Ginger extract	[79]
Leaf extract of *Pisoniagrandis*	[45]	Flower extract of *Cocciniagrandis*	[80]
Leaf extract*of Amaranthusdubius*	[46]	*Leaf extract of Vitexnegundo*	[36]
Leaf extract*of Elaeagnuslatifolia*	[47]	*Leaf extract of* Breyniaretusa	[81]
Leaf extract*of Hyacinthus orientalis*	[48]	*Leaf extract of* Costuspictus	[82]
Leaf extract*of Dianthus caryophyllus*	[48]	Flower extract of *Calendula officinalis*	[83]
Leaf extract of Euphorbia prostrata	[49]	Leaf extract of the medicinal plant *Adhatodavasica*	[84]
Grape fruit extract (*Vitisvinifera*)	[50]	Ficuselastica leaf extract	[85]
Leaf extract of Ocimum sanctum	[51]	T. procumbens fresh plant extract (leaf and stem)	[86]
Seed extract of P. granatum	[52]	Starfruit (*Averrhoa carambola*) extract	[87]
Leaf extract of Nerium oleander	[53]	Leaf extracts of *Abrusprecatorius*	[88]
Leaf extract of *Azhadirachtaindica*	[54]	*Achilleabiebersteinii* flowers extract	[89]
Leaf extract of Paederiafoetida	[55]	*Leaf extract of Musa balbisiana* (banana)	[90]
Leaf extract of *Adeniumobesum*	[56]	*Leaf extract of Azadirachtaindica* (neem)	[90]
Leaf extract*of Solanumtricobatum*	[57]	*Leaf extract of Ocimumtenuiflorum* (black tulsi).	[90]

Leaf extract*of Syzygiumcumini,*	[57]	*Catharanthusroseus* var. alba (*C. roseus* var. alba) callus extract	[91]
Leaf extract*of Centellaasiatica*	[57]	*Leaf extract of Curcuma longa*	[92]
Leaf extract*of Citrus sinensis*	[57]	*Calendula officinalis* petal extract	[93]
Leaf extract*of Artemisia nilagirica*	[58]	Leaf extract of Artemisia annua and Sidaacuta	[94]
Leaf extract*of Tinosporacordifolia*	[59]	Orange peel extract	[95]
Leaf extract of CeratoniaSiliqua	[60]	Emblicaofficinalis leafextract	[96]
Leaf extract of *Daturametel*	[61,62]	Leaf extract*of Azadirachtaindica*	[97]
Leaf extract of *Memecylonumbellatum.*	[63]	Indicum (L.) sweet leaf extract	[36]
Leaf extract of Coleus aromaticus	[64]	Aloe vera leaf extract	[37]
Leaf extract*of Rumexhymenosepalus*	[65]	Leaf extracts of *Anisomelesmalabarica*	[98]
Nigella sativa seeds extract	[66]	Young leaf aqueous extract of Sterculiafoetida	[99]
Leaf extract*of Jasminum grandiflorum*	[67]	Leaf extract of parsley (*Petroselinum crispum*)	[100]
Leaf extract*of Cymbopogoncitrullus*	[67]		

Applications of Ag-NPs

The progress of development and application of nano-colloids in medical science provides an entirely new scope for detecting various ailments [101]. The most explored applications of silver nanoparticles is in the use of wound dressings. The studies suggesting that dressings made of silver nanoparticle are highly beneficial in protecting the wound site from bacterial infection. Compared with other antimicrobial silver compounds, silver nanoparticles have better healing and cosmetics after healing properties.

Various groups conducted a prospective study to evaluate the use of silver nanoparticles dressing on a variety of chronic non-healing wounds. The study concluded that silver nanoparticles dressing have a beneficial effect of protecting the wound site from bacterial infection [102-104].

Mohajeri and his colleague conducted a study to determine the efficacy of nano particles coated leather in reducing the number of microorganisms which grow on diabetic foot ulcer and subsequently prevent diabetic foot infections [105]. The silver ions interact with 3 main components of the bacterial cell to produce a bactericidal effect: the peptidoglycan cell wall, the plasma membrane and bacterial (cytoplasmic) DNA.

Silver nanoparticles can be easily incorporated in cotton fabric and dressings and have significantly decreased wound-healing time [106]. Anti-inflammatory properties of silver nanoparticles also promote wound healing by reducing cytokine release, decreasing lymphocyte and mast cell infiltration [107-109]. Silver nanoparticles interact with sulfur-containing proteins of the bacterial membrane, as well as with phosphorus-containing compounds such as DNA to inhibit replication [110]. Nanoparticles are also useful in the dentistry specially in acryclic resin , tissue conditioners , dental adhsives , composites , dental cements , implants and maxillofacial prosthesis.

Kim and his colleague suggested that the antimicrobial mechanism of silver nanoparticles is related to the formation of free radicals and subsequent free radical–induces membrane damage [111]. The free radicals derived from the surface of silver nanoparticles are responsible for the antimicrobial activity. In a study by Miller and his colleague in 2010, the effect of nano-crystalline silver on the healing of leg ulcers was studied [112].

Summary and Future Prospects

This study reviewed several plant extracts for the synthesis of silver nanoparticles.The extract from different parts of plant was used for the reduction of Ag^+ ions. The reducing agents present in plant extracts reduced silver ions to nano-sized particles.The potential of silver nanoparticles for the wound healing in Diabetic foot ulcer patients were reviewed. These studies are of highly importance and would be useful in improving the treatment of Diabetic foot ulcer patients.

References

1. Chamberlain, R. C., Fleetwood, K., Wild, S. H., Colhoun, H. M., Lindsay, R. S., Petrie, J. R., ... & Leese, G. P. (2022). Foot Ulcer and Risk of Lower Limb Amputation or Death in People With Diabetes: A National Population-Based Retrospective Cohort Study. *Diabetes Care, 45*(1), 83-91.

2. Sharma, R., Sharma, S. K., Mudgal, S. K., Jelly, P., & Thakur, K. (2021). Efficacy of hyperbaric oxygen therapy for diabetic foot ulcer, a systematic review and meta-analysis of controlled clinical trials. *scientific reports, 11*(1), 1-12.

3. Asche, C., LaFleur, J., & Conner, C. (2011). A review of diabetes treatment adherence and the association with clinical and economic outcomes. *Clinical therapeutics, 33*(1), 74-109.

4. Kaul, K., Tarr, J. M., Ahmad, S. I., Kohner, E. M., & Chibber, R. (2013). Introduction to diabetes mellitus. *Diabetes,* 1-11.

5. Landon, M. B. (2009). Eunice Kennedy Shriver National Institute of Child Health and human development maternal-fetal medicine units network. A multicenter, randomized trial of treatment for mild gestational diabetes. *N Engl J Med, 361,* 1339-1348.

6. Zimmet, P., Alberti, K. G. M. M., & Shaw, J. (2001). Global and societal implications of the diabetes epidemic. *Nature, 414*(6865), 782-787.

7. Boulton, A. J., Vileikyte, L., Ragnarson-Tennvall, G., & Apelqvist, J. (2005). The global burden of diabetic foot disease. *The Lancet, 366*(9498), 1719-1724.

8. Amin, N., & Doupis, J. (2016). Diabetic foot disease: from the evaluation of the "foot at risk" to the novel diabetic ulcer treatment modalities. *World journal of diabetes, 7*(7), 153.

9. Lung, C. W., Wu, F. L., Liao, F., Pu, F., Fan, Y., & Jan, Y. K. (2020). Emerging technologies for the prevention and management of diabetic foot ulcers. *Journal of Tissue Viability, 29*(2), 61-68.

10. Liu, Y. X., Cao, Q. M., & Ma, B. C. (2019). Pathogens distribution and drug resistance in patients with acute cerebral infarction complicated with diabetes and nosocomial pulmonary infection. *BMC Infectious Diseases, 19*(1), 1-6.

11. Reiber, G. E., Vileikyte, L. O. R. E. T. T. A., Boyko, E. D., Del Aguila, M., Smith, D. G., Lavery, L. A., & Boulton, A. J. (1999). Causal pathways for incident lower-extremity ulcers in patients with diabetes from two settings. *Diabetes care, 22*(1), 157-162.
12. Lipsky, B. A., Berendt, A. R., Cornia, P. B., Pile, J. C., Peters, E. J., Armstrong, D. G., ... & Senneville, E. (2012). 2012 Infectious Diseases Society of America clinical practice guideline for the diagnosis and treatment of diabetic foot infections. *Clinical infectious diseases, 54*(12), e132-e173.
13. Boulton, A. J. (2007). Whither clinical research in diabetic sensorimotor peripheral neuropathy? Problems of end point selection for clinical trials. *Diabetes care, 30*(10), 2752-2753.
14. Pendsey, S. P. (2010). Understanding diabetic foot. *International journal of diabetes in developing countries, 30*(2), 75.
15. Memon, M. L., Ikram, M., Azhar, M., & Balouch, V. (2022). Comparison of efficacy of systemic antibiotics alone and combination of systemic antibiotics with gentamicin cream in diabetic foot infections. *Pakistan Journal of Medical Sciences, 38*(3).
16. Husain, M., & Agrawal, Y. O. (2022). Antimicrobial Remedies and Emerging Strategies for the Treatment of Diabetic Foot Ulcers. *Current Diabetes Reviews*.
17. Aminov, R. I. (2010). A brief history of the antibiotic era: lessons learned and challenges for the future. *Frontiers in microbiology, 1*, 134.
18. Dowd, S. E., Callaway, T. R., Wolcott, R. D., Sun, Y., McKeehan, T., Hagevoort, R. G., & Edrington, T. S. (2008). Evaluation of the bacterial diversity in the feces of cattle using 16S rDNA bacterial tag-encoded FLX amplicon pyrosequencing (bTEFAP). *BMC microbiology, 8*(1), 1-8.
19. Dowd, S. E., Sun, Y., Secor, P. R., Rhoads, D. D., Wolcott, B. M., James, G. A., & Wolcott, R. D. (2008). Survey of bacterial diversity in chronic wounds using pyrosequencing, DGGE, and full ribosome shotgun sequencing. *BMC microbiology, 8*(1), 1-15.

20. Laue, H., Schenk, A., Li, H., Lambertsen, L., Neu, T. R., Molin, S., & Ullrich, M. S. (2006). Contribution of alginate and levan production to biofilm formation by Pseudomonas syringae. *Microbiology*, *152*(10), 2909-2918.

21. Sim, W., Barnard, R. T., Blaskovich, M. A. T., & Ziora, Z. M. (2018). Antimicrobial silver in medicinal and consumer applications: a patent review of the past decade (2007–2017). *Antibiotics*, *7*(4), 93.

22. Maillard, J. Y., & Hartemann, P. (2013). Silver as an antimicrobial: facts and gaps in knowledge. *Critical reviews in microbiology*, *39*(4), 373-383.

23. Wright, J. B., Lam, K., & Burrell, R. E. (1998). Wound management in an era of increasing bacterial antibiotic resistance: a role for topical silver treatment. *American journal of infection control*, *26*(6), 572-577.

24. Behera, S., Debata, A., & Nayak, P. L. (2011). Biomedical applications of silver nanoparticles. *Journal of Asian Scientific Research*, *1*(1), 27-56.

25. Javed, B., Ikram, M., Farooq, F., Sultana, T., Mashwani, Z. U. R., & Raja, N. I. (2021). Biogenesis of silver nanoparticles to treat cancer, diabetes, and microbial infections: A mechanistic overview. *Applied Microbiology and Biotechnology*, *105*(6), 2261-2275.

26. Dhuldhaj, U., & Pandya, U. (2021). Combinatorial study of heavy metal and microbe interactions and resistance mechanism consort to microbial system. *Geomicrobiology Journal*, *38*(2), 181-189.

27. Demling, R. H., & DeSanti, L. (2001). Effects of silver on wound management. *Wounds*, *13*(1), 4-15.

28. de Moura, F. B. R., Ferreira, B. A., Muniz, E. H., Justino, A. B., Silva, A. G., de Azambuja Ribeiro, R. I. M., ... & Tomiosso, T. C. (2022). Antioxidant, anti-inflammatory, and wound healing effects of topical silver-doped zinc oxide and silver oxide nanocomposites. *International Journal of Pharmaceutics*, 121620.

29. Dulińska-Litewka, J., Dykas, K., Felkle, D., Karnas, K., Khachatryan, G., & Karewicz, A. (2021). Hyaluronic Acid-

Silver Nanocomposites and Their Biomedical Applications: A Review. *Materials*, *15*(1), 234.

30. Zhang, X. F., Liu, Z. G., Shen, W., & Gurunathan, S. (2016). Silver nanoparticles: synthesis, characterization, properties, applications, and therapeutic approaches. *International journal of molecular sciences*, *17*(9), 1534.

31. World Health Organization. (2019). *WHO global report on traditional and complementary medicine 2019*. World Health Organization.

32. Das, M., & Chatterjee, S. (2019). Green synthesis of metal/metal oxide nanoparticles toward biomedical applications: Boon or bane. In *Green synthesis, characterization and applications of nanoparticles* (pp. 265-301). Elsevier.

33. Majeed, M., Hakeem, K. R., & Rehman, R. U. (2022). Synergistic effect of plant extract coupled silver nanoparticles in various therapeutic applications-present insights and bottlenecks. *Chemosphere*, *288*, 132527.

34. Shekhawat, M. S., Kannan, N., & Manokari, M. (2012). Biogenesis of silver nanoparticles using leaf extract of Turnera ulmifolia Linn. and screening of their antimicrobial activity. *Journal of Ecobiotechnology*, *4*(1).

35. Christensen, L., Vivekanandhan, S., Misra, M., & Kumar Mohanty, A. (2011). Biosynthesis of silver nanoparticles using Murraya koenigii (curry leaf): an investigation on the effect of broth concentration in reduction mechanism and particle size. *Advanced Materials Letters*, *2*(6), 429-434.

36. Anandalakshmi, K., Venugobal, J., & Ramasamy, V. (2016). Characterization of silver nanoparticles by green synthesis method using Pedalium murex leaf extract and their antibacterial activity. *Applied nanoscience*, *6*(3), 399-408.

37. Vidhu, V. K., Aromal, S. A., & Philip, D. (2011). Green synthesis of silver nanoparticles using Macrotyloma uniflorum. *Spectrochimica Acta Part A: Molecular and Biomolecular Spectroscopy*, *83*(1), 392-397.

38. Banerjee, P., Satapathy, M., Mukhopahayay, A., & Das, P. (2014). Leaf extract mediated green synthesis of silver nanoparticles from widely available Indian plants: synthesis,

characterization, antimicrobial property and toxicity analysis. *Bioresources and Bioprocessing*, *1*(1), 1-10.

39. Akash Raj, S., Divya, S., Sindhu, S., Kasinthan, K., & Armugam, P. (2014). Studie s on synthesis, characterization and application of silver nanoparticle s using Mimosa pudica leaves. *Int J Pharm Pharm Sci*, *6*(2), 453-455.

40. Patel, R., Mahobia, N. K., Gendle, R., Kaushik, B., & Singh, S. K. (2010). Diuretic activity of leaves of Plectranthus amboinicus (Lour) Spreng in male albino rats. *Pharmacognosy research*, *2*(2), 86.

41. Prathna, T. C., Chandrasekaran, N., Raichur, A. M., & Mukherjee, A. (2011). Biomimetic synthesis of silver nanoparticles by Citrus limon (lemon) aqueous extract and theoretical prediction of particle size. *Colloids and Surfaces B: Biointerfaces*, *82*(1), 152-159.

42. Srivastava, R. (2014). A review on phytochemical, pharmacological, and pharmacognostical profile of Wrightia tinctoria: Adulterant of kurchi. *Pharmacognosy reviews*, *8*(15), 36.

43. Ahmad, N., Sharma, A. K., Sharma, S., Khan, I., Sharma, D. K., Shamsi, A., ... & Seervi, M. (2019). Biosynthesized composites of Au-Ag nanoparticles using Trapa peel extract induced ROS-mediated p53 independent apoptosis in cancer cells. *Drug and chemical toxicology*, *42*(1), 43-53.

44. Joseph, S., & Mathew, B. (2014). Microwave assisted biosynthesis of silver nanoparticles using the rhizome extract of Alpinia galanga and evaluation of their catalytic and antimicrobial activities. *Journal of Nanoparticles*, *2014*.

45. Gilaki, M. (2010). Biosynthesis of silver nanoparticles using plant extracts. *Journal of Biological sciences*, *10*(5), 465-467.

46. Iravani, S. (2014). Bacteria in nanoparticle synthesis: current status and future prospects. *International scholarly research notices*, *2014*.

47. Aritonang, H. F., Koleangan, H., & Wuntu, A. D. (2019). Synthesis of silver nanoparticles using aqueous extract of medicinal plants'(Impatiens balsamina and Lantana camara) fresh leaves and analysis of antimicrobial activity. *International journal of microbiology*, *2019*.

48. Mikhailova, E. O. (2021). Gold Nanoparticles: Biosynthesis and Potential of Biomedical Application. *Journal of Functional Biomaterials*, *12*(4), 70.

49. Castillo-Henríquez, L., Alfaro-Aguilar, K., Ugalde-Álvarez, J., Vega-Fernández, L., Montes de Oca-Vásquez, G., & Vega-Baudrit, J. R. (2020). Green synthesis of gold and silver nanoparticles from plant extracts and their possible applications as antimicrobial agents in the agricultural area. *Nanomaterials*, *10*(9), 1763.

50. Ahmad, S., Munir, S., Zeb, N., Ullah, A., Khan, B., Ali, J., ... & Ali, S. (2019). Green nanotechnology: A review on green synthesis of silver nanoparticles—An ecofriendly approach. *International journal of nanomedicine*, *14*, 5087.

51. Ramteke, C., Chakrabarti, T., Sarangi, B. K., & Pandey, R. A. (2013). Synthesis of silver nanoparticles from the aqueous extract of leaves of Ocimum sanctum for enhanced antibacterial activity. *Journal of chemistry*, *2013*.

52. Ibrahim, H. M. (2015). Green synthesis and characterization of silver nanoparticles using banana peel extract and their antimicrobial activity against representative microorganisms. *Journal of radiation research and applied sciences*, *8*(3), 265-275.

53. Krithiga, N., Rajalakshmi, A., & Jayachitra, A. (2015). Green synthesis of silver nanoparticles using leaf extracts of Clitoria ternatea and Solanum nigrum and study of its antibacterial effect against common nosocomial pathogens. *Journal of Nanoscience*, *2015*.

54. Lalitha, A., Subbaiya, R., & Ponmurugan, P. (2013). Green synthesis of silver nanoparticles from leaf extract Azhadirachta indica and to study its anti-bacterial and antioxidant property. *Int J Curr Microbiol App Sci*, *2*(6), 228-235.

55. Madhavaraj, L., Sethumadhavan, V. V., Geun, H. G., Mathur, N. K., & Si, W. K. (2013). Synthesis, characterization and evaluation of antimicrobial efficacy of silver nanoparticles using Paederia foetida leaf extract. *Intern. Res. J. Biol. Sci*, *15*(3), 76-80.

56. Farah, M. A., Ali, M. A., Chen, S. M., Li, Y., Al-Hemaid, F. M., Abou-Tarboush, F. M., ... & Lee, J. (2016). Silver nanoparticles synthesized from Adenium obesum leaf extract induced DNA damage, apoptosis and autophagy via generation of reactive oxygen species. *Colloids and Surfaces B: Biointerfaces, 141,* 158-169.

57. Banerjee, J., & Narendhirakannan, R. T. (2011). Biosynthesis of silver nanoparticles from Syzygium cumini (L.) seed extract and evaluation of their in vitro antioxidant activities. *Dig J Nanomater Biostruct, 6*(3), 961-968.

58. Rautela, A., & Rani, J. (2019). Green synthesis of silver nanoparticles from Tectona grandis seeds extract: characterization and mechanism of antimicrobial action on different microorganisms. *Journal of Analytical Science and Technology, 10*(1), 1-10.

59. Rajathi, K., Vijaya Raj, D., Anarkali, J., & Sridhar, S. (2012). Green Synthesis, characterization and in-vitro antibacterial activity of silver nanoparticles by using Tinospora cordifolia leaf extract. *Int J Green Chem Bioprocess, 2*(2), 15-19.

60. Devaraj, P., Kumari, P., Aarti, C., & Renganathan, A. (2013). Synthesis and characterization of silver nanoparticles using cannonball leaves and their cytotoxic activity against MCF-7 cell line. *Journal of nanotechnology, 2013.*

61. Gomathi, M., Rajkumar, P. V., Prakasam, A., & Ravichandran, K. (2017). Green synthesis of silver nanoparticles using Datura stramonium leaf extract and assessment of their antibacterial activity. *Resource-Efficient Technologies, 3*(3), 280-284.

62. Ojha, A. K., Rout, J., Behera, S., & Nayak, P. L. (2013). Green synthesis and characterization of zero valent silver nanoparticles from the leaf extract of Datura metel. *International journal of pharmaceutical research and allied sciences, 2*(1), 31-35.

63. Chandran, S. P., Chaudhary, M., Pasricha, R., Ahmad, A., & Sastry, M. (2006). Synthesis of gold nanotriangles and silver nanoparticles using Aloevera plant extract. *Biotechnology progress, 22*(2), 577-583.

64. Kuppusamy, P., Yusoff, M. M., Maniam, G. P., & Govindan, N. (2016). Biosynthesis of metallic nanoparticles using plant

derivatives and their new avenues in pharmacological applications–An updated report. *Saudi Pharmaceutical Journal*, *24*(4), 473-484.

65. Alvarez-Cirerol, F. J., López-Torres, M. A., Rodríguez-León, E., Rodríguez-Beas, C., Martínez-Higuera, A., Lara, H. H., ... & Iñiguez-Palomares, R. A. (2019). Silver nanoparticles synthesized with Rumex hymenosepalus: A strategy to combat early mortality syndrome (EMS) in a cultivated white shrimp. *Journal of Nanomaterials*, *2019*.

66. Widdatallah, M. O., Mohamed, A. A., Alrasheid, A. A., Widatallah, H. A., Yassin, L. F., Eltilib, S. H., & Ahmed, S. A. R. (2020). Green synthesis of silver nanoparticles using Nigella sativa seeds and evaluation of their antibacterial activity. *Advances in Nanoparticles*, *9*(2), 41-48.

67. Ashraf, J. M., Ansari, M. A., Fatma, S., Abdullah, S., Iqbal, J., Madkhali, A., ... & Ashraf, G. M. (2018). Inhibiting effect of zinc oxide nanoparticles on advanced glycation products and oxidative modifications: a potential tool to counteract oxidative stress in neurodegenerative diseases. *Molecular Neurobiology*, *55*(9), 7438-7452.

68. Panneerselvam, C., Murugan, K., Roni, M., Aziz, A. T., Suresh, U., Rajaganesh, R., ... & Benelli, G. (2016). Fern-synthesized nanoparticles in the fight against malaria: LC/MS analysis of Pteridium aquilinum leaf extract and biosynthesis of silver nanoparticles with high mosquitocidal and antiplasmodial activity. *Parasitology Research*, *115*(3), 997-1013.

69. Dangi, S., Gupta, A., Gupta, D. K., Singh, S., & Parajuli, N. (2020). Green synthesis of silver nanoparticles using aqueous root extract of Berberis asiatica and evaluation of their antibacterial activity. *Chemical Data Collections*, *28*, 100411.Ashok, D., Sandupatla, R., & Koyyati, R. (2017).

70. Phytomediated Synthesis of Silver Nanoparticles using Dicrostachys cinerea leaf extract and evaluation of its Antibacterial and Photo catalytic activity of Textile dye. *International Journal of ChemTech Research*, *10*, 302-314.

71. Francis, S., Joseph, S., Koshy, E. P., & Mathew, B. (2018). Microwave assisted green synthesis of silver nanoparticles using leaf extract of elephantopus scaber and its environmental and biological applications. *Artificial cells, nanomedicine, and biotechnology*, 46(4), 795-804.
72. Jemal, K., Sandeep, B. V., & Pola, S. (2017). Synthesis, characterization, and evaluation of the antibacterial activity of Allophylus serratus leaf and leaf derived callus extracts mediated silver nanoparticles. *Journal of Nanomaterials*, 2017.
73. Hafez, R. A., Abdel-Wahhab, M. A., Sehab, A. F., & Al-Zahraa, A. A. K. (2017). Green synthesis of silver nanoparticles using Morus nigra leave extract and evaluation their antifungal potency on phytopathogenic fungi. *J. Appl. Pharm. Sci*, 7, 041-048.
74. Nagati, V. B., Alwala, J., Koyyati, R., Donda, M. R., Banala, R., & Padigya, P. R. M. (2012). Green synthesis of plant-mediated silver nanoparticles using Withania somnifera leaf extract and evaluation of their antimicrobial activity. *Asian Pac J Trop Biomed*, 2, 1-5.
75. Waris, A., Din, M., Ali, A., Afridi, S., Baset, A., Khan, A. U., & Ali, M. (2021). Green fabrication of Co and Co3O4 nanoparticles and their biomedical applications: A review. *Open life sciences*, 16(1), 14-30.
76. Femi-Adepoju, A. G., Dada, A. O., Otun, K. O., Adepoju, A. O., & Fatoba, O. P. (2019). Green synthesis of silver nanoparticles using terrestrial fern (Gleichenia Pectinata (Willd.) C. Presl.): characterization and antimicrobial studies. *Heliyon*, 5(4), e01543.
77. Asimuddin, M., Shaik, M. R., Adil, S. F., Siddiqui, M. R. H., Alwarthan, A., Jamil, K., & Khan, M. (2020). Azadirachta indica based biosynthesis of silver nanoparticles and evaluation of their antibacterial and cytotoxic effects. *Journal of King Saud University-Science*, 32(1), 648-656.
78. Dada, A. O., Inyinbor, A. A., Idu, E. I., Bello, O. M., Oluyori, A. P., Adelani-Akande, T. A., ... & Dada, O. (2018). Effect of operational parameters, characterization and antibacterial studies

of green synthesis of silver nanoparticles using Tithonia diversifolia. *PeerJ*, *6*, e5865.

79. Alkhathlan, A. H., AL-Abdulkarim, H. A., Khan, M., Khan, M., AlDobiy, A., Alkholief, M., ... & Siddiqui, M. R. H. (2020). Ecofriendly Synthesis of Silver Nanoparticles Using Aqueous Extracts of Zingiber officinale (Ginger) and Nigella sativa L. Seeds (Black Cumin) and Comparison of Their Antibacterial Potential. *Sustainability*, *12*(24), 10523.

80. Manik, U. P., Nande, A., Raut, S., & Dhoble, S. J. (2020). Green synthesis of silver nanoparticles using plant leaf extraction of Artocarpus heterophylus and Azadirachta indica. *Results in Materials*, *6*, 100086.

81. Karunagaran, V. (2012). Biosynthesis of silver nanoparticles by microwave assisted method. National Conference on Recent Advances in Chemical and Environmental Engineering.

82. Aruna, A., Nandhini, R., Karthikeyan, V., & Bose, P. (2014). Synthesis and characterization of silver nanoparticles of insulin plant (costus pictus d. don) leaves. *Asian journal of biomedical and pharmaceutical sciences*, *4*(34), 1.

83. Raja, S., Ramesh, V., & Thivaharan, V. (2015). Antibacterial and anticoagulant activity of silver nanoparticles synthesised from a novel source–pods of Peltophorum pterocarpum. *Journal of Industrial and Engineering Chemistry*, *29*, 257-264.

84. Bhavyasree, P. G., & Xavier, T. S. (2020). Green synthesis of Copper Oxide/Carbon nanocomposites using the leaf extract of Adhatoda vasica Nees, their characterization and antimicrobial activity. *Heliyon*, *6*(2), e03323.

85. Sathishkumar, P., Vennila, K., Jayakumar, R., Yusoff, A. R. M., Hadibarata, T., & Palvannan, T. (2016). Phyto-synthesis of silver nanoparticles using Alternanthera tenella leaf extract: An effective inhibitor for the migration of human breast adenocarcinoma (MCF-7) cells. *Bioprocess and biosystems engineering*, *39*(4), 651-659.

86. Mahiuddin, M., Saha, P., & Ochiai, B. (2020). Green synthesis and catalytic activity of silver nanoparticles based on piper chaba stem extracts. *Nanomaterials*, *10*(9), 1777.

87. Al-Otibi, F., Perveen, K., Al-Saif, N. A., Alharbi, R. I., Bokhari, N. A., Albasher, G., ... & Al-Mosa, M. A. (2021). Biosynthesis of silver nanoparticles using Malva parviflora and their antifungal activity. *Saudi Journal of Biological Sciences*, *28*(4), 2229-2235.

88. Al-Shmgani, H. S., Mohammed, W. H., Sulaiman, G. M., & Saadoon, A. H. (2017). Biosynthesis of silver nanoparticles from Catharanthus roseus leaf extract and assessing their antioxidant, antimicrobial, and wound-healing activities. *Artificial cells, nanomedicine, and biotechnology*, *45*(6), 1234-1240.

89. Fafal, T. U. G. C. E., Taştan, P. E. L. I. N., Tüzün, B. S., Ozyazici, M., & Kivcak, B. (2017). Synthesis, characterization and studies on antioxidant activity of silver nanoparticles using Asphodelus aestivus Brot. aerial part extract. *South African Journal of Botany*, *112*, 346-353.

90. Tailor, G., Yadav, B. L., Chaudhary, J., Joshi, M., & Suvalka, C. (2020). Green synthesis of silver nanoparticles using Ocimum canum and their anti-bacterial activity. *Biochemistry and Biophysics Reports*, *24*, 100848.

91. Mukunthan, K. S., Elumalai, E. K., Patel, T. N., & Murty, V. R. (2011). Catharanthus roseus: a natural source for the synthesis of silver nanoparticles. *Asian pacific journal of tropical biomedicine*, *1*(4), 270-274.

92. Shameli, K., Ahmad, M. B., Zamanian, A., Sangpour, P., Shabanzadeh, P., Abdollahi, Y., & Zargar, M. (2012). Green biosynthesis of silver nanoparticles using Curcuma longa tuber powder. *International journal of nanomedicine*, *7*, 5603.

93. Birusanti, A. B., Mallavarapu, U., Nayakanti, D., Espenti, C. S., & Mala, S. (2019). Sustainable green synthesis of silver nanoparticles by using Rangoon creeper leaves extract and their spectral analysis and anti-bacterial studies. *IET nanobiotechnology*, *13*(1), 71-76.

94. Idrees, M., Batool, S., Kalsoom, T., Raina, S., Sharif, H. M. A., & Yasmeen, S. (2019). Biosynthesis of silver nanoparticles using Sida acuta extract for antimicrobial actions and corrosion inhibition potential. *Environmental technology*, *40*(8), 1071-1078.

95. Sharma, A., Wakode, S., Sharma, S., Fayaz, F., & Pottoo, F. H. (2020). Methods and strategies used in green chemistry: a review. *Current Organic Chemistry*, *24*(22), 2555-2565.

96. Ponarulselvam, S., Panneerselvam, C., Murugan, K., Aarthi, N., Kalimuthu, K., & Thangamani, S. (2012). Synthesis of silver nanoparticles using leaves of Catharanthus roseus Linn. G. Don and their antiplasmodial activities. *Asian Pacific journal of tropical biomedicine*, *2*(7), 574-580.

97. Ahmed, S., Ahmad, M., Swami, B. L., & Ikram, S. (2016). A review on plants extract mediated synthesis of silver nanoparticles for antimicrobial applications: a green expertise. *Journal of advanced research*, *7*(1), 17-28.

98. Govindarajan, M., & Benelli, G. (2017). A facile one-pot synthesis of eco-friendly nanoparticles using Carissa carandas: ovicidal and larvicidal potential on malaria, dengue and filariasis mosquito vectors. *Journal of cluster science*, *28*(1), 15-36.

99. Anh, N. P., Mi, T. T. A., Linh, D. H. T., Van, N. T. T., Cuong, H. T., Van Minh, N., & Tri, N. (2018). Fortunella japonica extract as a reducing agent for green synthesis of silver nanoparticles. *International Journal of Engineering & Technology*, *7*(3), 1570-1575.

100. Khan, S. U., Anjum, S. I., Ansari, M. J., Khan, M. H. U., Kamal, S., Rahman, K., ... & Khan, D. (2019). Antimicrobial potentials of medicinal plant's extract and their derived silver nanoparticles: A focus on honey bee pathogen. *Saudi journal of biological sciences*, *26*(7), 1815-1834.

101. Tehri, N., Vashishth, A., Gahlaut, A., & Hooda, V. (2022). Biosynthesis, antimicrobial spectra and applications of silver nanoparticles: Current progress and future prospects. *Inorganic and Nano-Metal Chemistry*, *52*(1), 1-19.

102. Agrawal, S., Bhatt, M., Rai, S. K., Bhatt, A., Dangwal, P., & Agrawal, P. K. (2018). Silver nanoparticles and its potential applications: A review. *Journal of Pharmacognosy and Phytochemistry*, *7*(2), 930-937.

103. Naidu Krishna, S., Govender, P., & Adam, J. K. (2015). Nano silver particles in biomedical and clinical applications. *Journal of pure and applied microbiology (Print)*.

104. Liang, Y., Liang, Y., Zhang, H., & Guo, B. (2022). Antibacterial biomaterials for skin wound dressing. *Asian Journal of Pharmaceutical Sciences*.

105. Aalaa, M., Malazy, O. T., Sanjari, M., Peimani, M., & Mohajeri-Tehrani, M. R. (2012). Nurses' role in diabetic foot prevention and care; a review. *Journal of Diabetes & Metabolic Disorders*, *11*(1), 1-6.

106. Ovais, M., Ahmad, I., Khalil, A. T., Mukherjee, S., Javed, R., Ayaz, M., ... & Shinwari, Z. K. (2018). Wound healing applications of biogenic colloidal silver and gold nanoparticles: recent trends and future prospects. *Applied microbiology and biotechnology*, *102*(10), 4305-4318.

107. Saxena, N. (2020). In vivo applications of green synthesized silver nanoparticles as hydrogel dressings.

108. Chi, N. T. L., Narayanan, M., Chinnathambi, A., Govindasamy, C., Subramani, B., Brindhadevi, K., ... & Pikulkaew, S. (2022). Fabrication, characterization, anti-inflammatory, and anti-diabetic activity of silver nanoparticles synthesized from Azadirachta indica kernel aqueous extract. *Environmental research*, 112684.

109. Negi, A., Vishwakarma, R. K., & Negi, D. S. (2022). Synthesis and evaluation of antibacterial, anti-fungal, anti-inflammatory properties of silver nanoparticles mediated via roots of Smilax aspera. *Materials Today: Proceedings*.

110. Khalandi, B., Asadi, N., Milani, M., Davaran, S., Abadi, A. J. N., Abasi, E., & Akbarzadeh, A. (2017). A review on potential role of silver nanoparticles and possible mechanisms of their actions on bacteria. *Drug research*, *11*(02), 70-76.

111. Im, A. R., Han, L., Kim, E. R., Kim, J., Kim, Y. S., & Park, Y. (2012). Enhanced antibacterial activities of Leonuri herba extracts containing silver nanoparticles. *Phytotherapy research*, *26*(8), 1249-1255.

112. Miller, C. N., Newall, N., Kapp, S. E., Lewin, G., Karimi, L., Carville, K., ... & Santamaria, N. M. (2010). A randomized-controlled trial comparing cadexomer iodine and nanocrystalline silver on the healing of leg ulcers. *Wound repair and regeneration*, *18*(4), 359-367.

[*1]Professor,

Department of Chemistry,

Poddar International College , Mansarovar ,

Jaipur, Rajasthan, India

email : drmeenumangal@gmailcom

[2]Prosthodontist and Dental Consultant,

Agrawal Dental Hospital and Research Centre,

Jawahar Nagar, Jaipur,Rajasthan, India

email : drsunilmangal@gmail.com

12. Toxic Effect of Malathion on Hb Content of *Clarius Batrachus*

R. S. Magar

Abstract

The freshwater fish *Clarius batrachus* was exposed to Sublethal Concentration of 0.05,0.25, and 0.5 ppm organophosphate pesticide, Malathion for 7, 14, 21 & 28 days.The Hb concertation decreases as the concertation of pesticide increase.Chronic effect of pesticide also noticed during duration, as duration increases Hb concertation also decreases.

Keywords : *Clarius batrachus*, Malathion, pesticide, chronic.

Introduction

Increased use of pesticide in most tropical countries has been reported to results in severe toxicities and bioaccumulation

Malathion is commonly used organo phosphorous pesticide. While most of the malathion will stay inthe areas where it is applied, some can move to areas away from where it was applied by rain, fog and wind. Once malathion is introduced into the evironment, it may cause serious intimidation to aquatic organisms and is notorious to cause severe metabolic disturbances in non target species like fish and fresh water mussels. USEPA (2005)

Clarius batrachus (Family- claride) is a common food fish in India scanty research is found onfresh water fish abundantly present in local river Godavari Dist. Nanded. It is one of the major source of food of poor population in local area. The present study was designed to study impact of sublethal concentration of of malathion (0.05, ,0.25&0.5)on Hb concertation in fresh water fish *Clarius batrachus* during exposure period of 7,14,21 and 28 Days.

Material and Methods :

For present study, commercial grade malathion (50% manufactured by Coromandal fertilizer limited, Coromandal house, pesticide division, Ranipet, Veilore (TN), India) was procured from the local market. Healthy specimens of *Clarius batrachus* were collected from local river Godavari Dist. Nanded. The average length and wet weight (24.5± 2cm) and weight 158±4gm) respectively. Fishes were treated with 0.1 % KmNo₄ solution for 2 min. to avoid any dermal infection. The fish stock was then maintained in 100 liter glass aquaria for 14 days to acclimatize under laboratory condition. The fishes were fed with pieces of live earth warm on alternate days. A stock solution was prepared in acetone and mixed in water to obtain required dilutions. The LC50 value for 96 hours of malathion was determined by procedure of Finney (1971).The LC50 of malathion for 96 hours for *Clarius batrachus* was 5 mg/liter. Fishes were exposed to sub lethal concentration (0.05,0.25&0.5 ppm) of malathion, simultaneously control group was also maintained. Haemoglobin standard Sahli's method outlined by Wintobe(1968).

Results :

In the present investigation the protein content at control 13.5 compare with experiment in 7, 14, 21 and 28dayswas 10.28, 9.17, 8.21 and 6.71 mg/dl in for concertation 0.05.For Concertation 0.25 treated values dresses from 9.28 to 5.23 . For concertation 0.25 values decreases from 8.13 to 4.25for duration 7 days to 28 days. Changes in Hb concertation of of *Clarius batrachus* is presented in Table 1.

Discussion :

In *Clarius batrachus* decreasing trend in Hb Concertation on exposure to Malathion at different chronic hours (7 days to 28 days) has been observed.The decreasing trend has been reported by Thakur and Sahai (1968) in *Channa punctatus* exposed to BHC; Nath and Banerjee (1995) in *Heteropneustes fossils.*Kumar *et al.*(2006) in *Clarius batrachus* ; Sexena and Seth(2002) in *Channa punctatus* after cypermethrin treatment in *Channa punctatus.* Reduction in Hb. Congratulations might be due to RBC count reduction during exposed hours leads to hypohaemoglobinemia.

Table. 1 Haemoglobin Concentration (mg/dl) in *Clarius batrachus* exposed to Malathion

Concentration of Malathion in ppm	Duration in days 7	14days	21days	28days
control	13.5± 0.3	13.5± 0.2	13.5± 0.3	13.5± 0.3
Experimental 0.05 ppm	10.28 ± 0.4	9.17 ± 0.5	8.21± 0.2	6.71±0.3
0.25ppn	9.28± 0.2	7.22±0.3	7.01 ± 0.05	5.23 ± 0.4
0.5ppm	8.13 ± 0.4	6.23 ± 0.4	5.11 ± 0.3	4.25 ± 0.2

(Values are mean SD of six replicates, * P<0.05, * P <0.01, ** P>0.01, significant when student's test was applied between control and experimental groups)

Conclusion :

In the present investigation the effect of organophosphate malathion on the Hb concertation in *Clarius batrachus*has been studied. Decreasing trend has been noticed during exposed hours.

Acknowledgement :

The author is thankful to Principal Shri Datta A.C.& S.College Hadgaonfor providing the laboratory facilities during this work.

References :

Kumar,Y .,Malik,M.,panwar,A.& Singh,H.S.(2005)- Efficacy of clove oil as an anesthetic for freshwater fish- *Clarius batrachus (Linn.)* J. Exp. Zool. India,8: 225-234.

Nath, R . Banerjee, V .(1995)- Sub-lethal effect of devithion on the haematological parameters in *Heteropneustes* fossilis(Bloch.) J. Fresh Water Biol.,7 (4):261-264.

USEPA (2005) United states environmental protection agency.

Sexena,K. K.& Seth,N .(2002) – Toxic effect of cypermethrin on certain haematological aspects of fresh water fish *Channa punctatus.* Bull.Environ. Contam. Toxicol., 69: 364- 369

Thakur N and Sahai, S. (1987)- Carbaryl induced haematological alterations in teleost, *Garra gotyla gotyla (Gray)* In: Environ.and Ecotoxicol.

Department of Zoology,
Shri Datta A.C.& S.College Hadgaon
Nanded, Maharashtra, India.
email : rajendra.magar0999@gmail.com

13. The Future of Sustainable Agriculture with Microbial Solutions

Ishan Tiwari

Abstract

This book chapter explores the potential of microbial solutions in shaping the future of sustainable agriculture. Microorganisms have been recognized as essential players in maintaining soil health and crop productivity, and recent advancements in microbiology and biotechnology have expanded the possibilities for harnessing their potential for sustainable agriculture. The chapter discusses the current challenges facing agriculture, including the need for increased food production to feed a growing population, environmental degradation caused by conventional farming practices, and the emerging threat of climate change. It then explores the potential of microbial solutions, such as biofertilizers, biopesticides, and bio stimulants, in addressing these challenges. The chapter also discusses the latest research and development in the field of microbial solutions and highlights the potential of new technologies, such as synthetic biology, for creating customized microbial solutions for specific crops and farming systems. Ultimately, the chapter argues that microbial solutions have the potential to revolutionize sustainable agriculture and pave the way for a more resilient and productive food system.

Keywords: Microbiology, biotechnology, sustainability, agriculture, productive

Introduction

Agriculture is a vital sector that supports the livelihoods of millions of people worldwide. It is responsible for feeding the growing global population, providing income to farmers, and contributing to the economy. However, conventional agricultural practices, such as intensive monoculture farming and the use of chemical fertilizers and pesticides, have led to environmental degradation, soil erosion, and the loss of biodiversity. Furthermore, agriculture is a significant contributor to greenhouse gas emissions, which contribute to climate change (FAO, 2019).

The challenges facing agriculture are compounded by a rapidly growing global population, which is projected to reach 9.7 billion by 2050 (UN, 2019). To meet this demand, global food production must increase by 70% (FAO, 2018). However, traditional farming methods are not sustainable, and there is an urgent need to develop new and innovative approaches that can ensure food security while minimizing environmental impact.

Microorganisms have been recognized as essential players in maintaining soil health and crop productivity, and recent advancements in microbiology and biotechnology have expanded the possibilities for harnessing their potential for sustainable agriculture. Microbial solutions are defined as products or technologies that utilize microorganisms to improve plant health, increase crop yields, and reduce the use of synthetic fertilizers and pesticides (Bhattacharyya et al., 2019).

Sustainability in agriculture involves the long-term maintenance of soil productivity using natural resources without degrading the environment. Essential to achieving sustainable agriculture is the preservation of natural resources, including diverse and functional microbial populations in the soil. Integrated soil management has gained acceptance among environmentalists and emphasizes the management of ecosystem functioning through nutrient cycling, waste management, and the management of soil microbial diversity. Microorganisms are the most abundant and diverse natural resources on earth, but are often overlooked due to their small size. The estimated total number of microorganisms on earth is in the range of 4-6x1030 cells, and they add the same amount of carbon, nitrogen, and phosphorus to the earth as terrestrial plants (Forney et al., 2004). However, only a few culturable microorganisms have been used in agriculture as biofertilizers, biocontrol agents, and bioremediation agents, while vast pools of unculturable microorganisms remain unexplored. Microorganisms function not in isolation, but as a community, and therefore, it is crucial to manage microbial communities instead of mass-producing and applying a single microorganism in the soil for sustainable agriculture. Recently, attempts have been made to include microbial diversity as a component of soil classification and soil management programs

(Rao and Patra, 2009). The important question is whether sustainable agricultural production can be achieved through the management of the microbial component of the soil.

Challenges Faced by Agriculture Sector

Agriculture is a crucial sector for economic growth and development, and it relies heavily on soil health and productivity. However, farmers around the world are facing numerous challenges in maintaining the health and productivity of their agricultural soil. Some of the problems faced by farmers include soil degradation, soil erosion, nutrient depletion, soil compaction, and soil pollution. These problems are often caused by intensive farming practices, inappropriate use of chemical fertilizers and pesticides, land-use changes, and climate change.

Soil degradation is a major concern for farmers, as it affects the physical, chemical, and biological properties of the soil. Soil erosion is another problem that leads to loss of topsoil, reduced water-holding capacity, and decreased soil fertility. Nutrient depletion is also a significant issue, as continuous cropping without proper soil management can result in soil nutrient imbalances and deficiencies. Soil compaction is a common problem, especially in areas with heavy machinery and livestock, which can lead to reduced water infiltration and root growth.

Soil pollution is another major challenge that farmers face, as it affects both soil and crop quality. Pollution can result from the inappropriate use of agrochemicals, industrial activities, and waste disposal. The accumulation of heavy metals, pesticides, and other toxic substances in the soil can lead to decreased soil fertility, reduced crop yields, and health hazards for humans and animals. Soil pollution can result from the use of chemical fertilizers and pesticides, improper disposal of industrial waste, and other human activities (Pimentel et al., 2005). Exposure to contaminated soil can lead to serious health problems, such as cancer, birth defects, and developmental disorders (Khan et al., 2017). Therefore, it is crucial to adopt sustainable soil management practices that minimize the use of chemicals and promote organic farming methods.

Furthermore, the degradation of soil biodiversity is another significant challenge for farmers. Soil biodiversity refers to the diverse range of microorganisms, plants, and animals that live in the soil and contribute to its fertility and productivity (Bardgett and van der Putten, 2014). The loss of soil biodiversity can lead to decreased soil fertility, reduced crop yields, and increased susceptibility to pests and diseases. Therefore, promoting soil biodiversity is an essential component of sustainable agriculture.

Addressing these problems requires a shift towards more sustainable farming practices that focus on maintaining and improving soil health. This includes the use of organic fertilizers, conservation tillage, crop rotation, cover crops, and integrated pest management. It also involves the use of modern technologies such as precision agriculture, remote sensing, and artificial intelligence to optimize soil management practices. In addition to the challenges mentioned above, farmers also face issues related to soil salinity, acidity, and alkalinity. Soil salinity is a result of excess salts in the soil, which can lead to reduced water uptake by plants and decreased crop yields. Soil acidity and alkalinity, on the other hand, can affect the availability of essential nutrients in the soil, leading to nutrient deficiencies in crops.

Climate change is another significant challenge for farmers, as it affects soil health and productivity through changes in temperature, rainfall patterns, and extreme weather events. Climate change can also exacerbate other soil-related problems, such as soil erosion and degradation.

The problems faced by farmers on agricultural soil not only affect their livelihoods but also have wider implications for food security, biodiversity, and the environment. Therefore, it is crucial to adopt a holistic approach to address these challenges and promote sustainable agriculture.

Efforts are being made to address these challenges through various initiatives, such as the Global Soil Partnership, the Sustainable Development Goals, and the Climate Smart Agriculture approach. These initiatives aim to promote sustainable soil management practices, increase awareness about soil health, and enhance research and innovation in the field of soil science.

Microbial Solutions for Sustainable Agriculture

Soil microbial diversity is essential for ecosystem functioning, as it drives nutrient cycling, organic matter decomposition, and soil structure formation. Microorganisms also play a vital role in plant growth and health by providing essential nutrients, producing plant growth-promoting substances, and suppressing plant pathogens. Studies have shown that microbial diversity is positively correlated with soil health, crop productivity, and resilience to environmental stress (Banerjee et al., 2016; Liang et al., 2017).

Microbial Inoculants in Agriculture

Microbial inoculants are defined as live microorganisms that are added to soil or plant systems to enhance plant growth, health, and yield. Microbial inoculants include bacteria, fungi, and other microorganisms that have beneficial effects on plants, either directly or indirectly. There are various types of microbial inoculants, including biofertilizers, biocontrol agents, and soil amendments.

Biofertilizers are microbial inoculants that improve soil fertility and plant nutrition by fixing atmospheric nitrogen, solubilizing phosphorus, and producing other essential nutrients. Biofertilizers include nitrogen-fixing bacteria, such as Rhizobia and Azotobacter, and phosphorus-solubilizing bacteria, such as Bacillus and Pseudomonas. Biofertilizers have been shown to increase crop yields and reduce fertilizer use, leading to cost savings and environmental benefits (Vessey, 2003; Mohammadi et al., 2017).

Biocontrol agents are microbial inoculants that suppress plant pathogens and pests by various mechanisms, such as competition, predation, and production of antimicrobial compounds. Biocontrol agents include fungi, bacteria, and viruses that are specific to certain plant pathogens or pests. Biocontrol agents have been used successfully in controlling plant diseases, such as Fusarium wilt and root rot, and reducing pesticide use (Compant et al., 2005; Kloepper et al., 2004).

Soil amendments are microbial inoculants that improve soil physical and chemical properties, such as soil structure, pH, and organic matter content. Soil amendments include compost, vermicompost, and other organic matter sources that provide nutrients and support microbial growth. Soil amendments have been shown to improve soil fertility, water-holding capacity, and crop productivity (Subler et al., 2014; Raza et al., 2017).

Soil microbial diversity is also critical in nutrient cycling, which is essential for maintaining soil fertility and productivity. Microorganisms play a significant role in breaking down organic matter, releasing nutrients such as nitrogen, phosphorus, and sulfur back into the soil, which plants can then use for their growth and development (Philippot et al., 2013). The microbial-mediated process of mineralization, nitrification, and denitrification is essential for the nitrogen cycle, a critical component of plant nutrition. In addition, soil microbes also play a critical role in the phosphorus cycle by solubilizing inorganic phosphorus in the soil, making it available to plants (Richardson et al., 2009).

However, agricultural practices such as excessive tillage, mono-cropping, and the use of synthetic fertilizers and pesticides can harm soil microbial diversity, causing a decline in soil health and productivity (Hartman et al., 2018). For example, the use of synthetic fertilizers can lead to a decline in the diversity and abundance of soil microorganisms due to their toxic effects on soil microorganisms (Jangid et al., 2008). Similarly, the use of pesticides can also have negative impacts on soil microbial diversity and function, leading to imbalances in the microbial community structure and potential reductions in soil fertility and productivity (Berg and Smalla, 2009).

The use of microbial solutions in sustainable agriculture can mitigate the negative impacts of conventional agricultural practices on soil microbial diversity and function. Microbial solutions include microbial inoculants, biofertilizers, and biopesticides, which are composed of specific microbial strains or communities that can improve soil health and plant growth while reducing the need for synthetic fertilizers and pesticides (Berg and Smalla, 2009).

Microbial inoculants are formulated mixtures of beneficial microorganisms, including bacteria, fungi, and mycorrhizae, which are applied to the soil to promote plant growth and health. They can enhance nutrient availability and uptake, improve soil structure and water-holding capacity, and suppress soil-borne pathogens (Bashan and de-Bashan, 2010). Biofertilizers are microbial-based fertilizers that contain beneficial microorganisms, such as nitrogen-fixing bacteria, which can convert atmospheric nitrogen into plant-available forms (Bhattacharyya and Jha, 2012). Biopesticides are microbial-based products that can control pests and diseases, such as Bacillus thuringiensis (Bt), which produces toxins that are lethal to specific pests (Kumar et al., 2016).

Microbial solutions offer several benefits for sustainable agriculture. They can reduce the need for synthetic fertilizers and pesticides, which can lead to cost savings for farmers and reduce the negative impacts of these inputs on soil health and environmental quality (Berg and Smalla, 2009). They can also improve soil health and productivity by enhancing soil structure, water-holding capacity, and nutrient availability (Bashan and de-Bashan, 2010). Furthermore, microbial solutions can promote plant growth and development, leading to higher yields and better-quality crops (Bhattacharyya and Jha, 2012).

In recent years, there has been growing interest in the use of microbial solutions for sustainable agriculture. Several studies have demonstrated the positive effects of microbial inoculants, biofertilizers, and biopesticides on crop yield, quality, and soil health (Gopalakrishnan et al., 2015; Verma et al., 2019). However, the effectiveness of microbial solutions can depend on several factors, such as soil type, crop species, and environmental conditions, which can affect microbial survival and activity (Bhattacharyya and Jha, 2012)

Advancements in Microbial Solutions

Microbial inoculants have been extensively studied and applied in agriculture for their ability to improve soil health, plant growth, and disease resistance. Recent advancements in microbial inoculants have focused on the identification of novel microorganisms with specific functions and the development of improved formulations that enhance microbial survival and activity.

One of the recent advancements in microbial inoculants is the identification of plant growth-promoting rhizobacteria (PGPR) with multiple beneficial functions. PGPR are a group of beneficial bacteria that colonize plant roots and promote plant growth through various mechanisms, such as nutrient acquisition, hormone production, and disease suppression. Recent studies have identified novel PGPR strains with multiple beneficial functions, such as Bacillus amyloliquefaciens and Bacillus subtilis, which can enhance plant growth, nutrient uptake, and disease resistance (Kumar et al., 2020). These strains have been formulated into microbial inoculants and applied to various crops, such as wheat, maize, and soybean, with positive effects on yield and quality (Zhang et al., 2018).

Another recent advancement in microbial inoculants is the development of improved formulations that enhance microbial survival and activity in the soil. Traditional microbial inoculants often suffer from low survival rates and limited efficacy due to various environmental stresses, such as drought, heat, and competition from other microorganisms. Recent developments in formulation technology have focused on improving microbial survival and activity by encapsulating microorganisms in protective matrices, such as alginate beads, and by adding various growth-promoting substances, such as humic acid and amino acids, to the formulation (Huang et al., 2021). These improved formulations have shown higher microbial survival rates and improved plant growth-promoting effects in field trials (Tiwari et al., 2021).

Another promising development in microbial solutions is the use of plant growth-promoting rhizobacteria (PGPR). These are bacteria that colonize the roots of plants and promote their growth and development through various mechanisms such as nitrogen fixation, production of growth hormones, and biocontrol of plant pathogens (Ahmad et al., 2011). Some examples of PGPR include Azospirillum, Bacillus, Pseudomonas, and Rhizobium.

One study by Bashan and de-Bashan (2010) demonstrated the effectiveness of PGPR in enhancing plant growth and nutrient uptake in different crops such as wheat, maize, and tomato. The authors reported that the inoculation of soil with Azospirillum and other PGPR increased plant biomass, chlorophyll content, and nutrient uptake compared to control plants. In addition, PGPR inoculation also improved soil quality by increasing soil organic matter, nitrogen, and phosphorus content.

Another promising area of research in microbial solutions is the use of mycorrhizal fungi. These are fungi that form a mutualistic relationship with plant roots, enhancing nutrient uptake and plant growth. Mycorrhizal fungi can solubilize phosphorus and other nutrients, making them available to plants, and can also improve soil structure and water-holding capacity (Smith and Read, 2008).

A study by Gopalakrishnan et al. (2015) demonstrated the effectiveness of mycorrhizal fungi in improving the growth and yield of different crops such as rice, maize, and soybean. The authors reported that the inoculation of soil with mycorrhizal fungi increased plant biomass, root length, and nutrient uptake compared to control plants. In addition, mycorrhizal inoculation also improved soil quality by increasing soil organic matter and nitrogen content.

Microbial solutions also offer potential benefits for organic farming, where the use of synthetic fertilizers and pesticides is restricted. The use of microbial-based fertilizers and biocontrol agents can provide an effective alternative for organic farmers to maintain soil health and manage pests and diseases (Köhl et al., 2019).

A study by Verma et al. (2019) demonstrated the effectiveness of biocontrol agents in controlling plant diseases in organic farming. The authors reported that the use of biocontrol agents such as Trichoderma and Pseudomonas reduced the incidence of plant diseases such as Fusarium wilt and powdery mildew in different crops such as tomato, cucumber, and chili pepper. In addition, biocontrol agents also improved plant growth and yield compared to control plants.

Despite the potential benefits of microbial solutions in agriculture, there are also some challenges to their widespread adoption. One of the main challenges is the lack of standardized testing and regulation of microbial products (Berg and Smalla, 2009). This can lead to uncertainty about the effectiveness and safety of microbial products and can limit their use in conventional agriculture.

Another challenge is the need for more research on the interactions between microbial communities and their host plants (Philippot et al., 2013). Microbial communities are complex and dynamic, and their interactions with plant roots and other soil organisms can be affected by various factors such as soil type, crop species, and environmental conditions. More research is needed to understand these interactions and to develop microbial solutions that are tailored to specific crops and soil conditions.

Synthetic Biology : Customized Microbial Solutions for Sustainable Agriculture

Synthetic biology is an emerging field of science that involves the engineering of biological systems for specific purposes. In agriculture, synthetic biology has the potential to provide customized microbial solutions that can improve soil health and plant growth while reducing the need for synthetic fertilizers and pesticides. This paragraph will explore the potential of synthetic biology in agriculture and discuss some of the challenges that need to be addressed to realize its full potential.

One of the main goals of synthetic biology in agriculture is to develop customized microbial solutions that can improve soil health and plant growth. This can be achieved by engineering microbes that can perform specific functions such as nutrient cycling, disease suppression, and stress tolerance. For example, scientists can engineer microbes that can produce specific enzymes that can break down organic matter and release nutrients such as nitrogen and phosphorus. This can reduce the need for synthetic fertilizers and improve soil health by increasing the availability of nutrients to plants. Similarly, scientists can engineer microbes that can produce specific compounds that can suppress plant pathogens and pests. This can reduce the need for synthetic pesticides and improve plant health by reducing the incidence of diseases and pests. Finally, scientists can engineer microbes that can tolerate specific environmental stresses such as drought and high salinity. This can improve plant growth and yield in regions where these stresses are common.

One of the challenges of developing customized microbial solutions is the complexity of microbial communities. Microbial communities are complex and dynamic, and their interactions with plants and other soil organisms can be affected by various factors such as soil type, crop species, and environmental conditions. To develop effective microbial solutions, scientists need to understand the interactions between microbes and their host plants and other soil organisms. This requires a deep understanding of microbial ecology and soil biology.

Another challenge of developing customized microbial solutions is the need for efficient and reliable genetic tools for engineering microbes. Synthetic biology involves the engineering of biological systems at the genetic level. To engineer microbes, scientists need efficient and reliable genetic tools that can manipulate genes and other genetic elements. While there have been significant advancements in genetic engineering tools, there is still a need for more efficient and reliable tools that can be used for a wide range of microbial species.

Despite these challenges, there have been significant advancements in synthetic biology for agriculture. For example, scientists have engineered microbes that can produce plant growth-promoting compounds such as auxins and cytokinins (Glick, 2012). These compounds can stimulate plant growth and increase yield. Similarly, scientists have engineered microbes that can produce enzymes that can break down organic matter and release nutrients such as nitrogen and phosphorus (Kumar et al., 2019). These microbes can reduce the need for synthetic fertilizers and improve soil health. Finally, scientists have engineered microbes that can tolerate specific environmental stresses such as drought and high salinity (Li et al., 2017). These microbes can improve plant growth and yield in regions where these stresses are common.

In addition to these advancements, there are also ongoing efforts to develop new genetic tools for engineering microbes. For example, scientists are developing new CRISPR-Cas systems that can be used to manipulate genes in a wide range of microbial species (Makarova et al., 2020). These new tools can improve the efficiency and reliability of genetic engineering and expand the range of microbial species that can be engineered.

To realize the full potential of synthetic biology in agriculture, there is a need for collaboration between scientists, farmers, and policymakers. Scientists need to work closely with farmers to develop microbial solutions that are tailored to specific crops and soil conditions. Policymakers need to create a regulatory environment that promotes the safe and responsible use of synthetic biology in agriculture. Finally, farmers need to be educated about the potential benefits and risks of synthetic biology and be given the tools and resources to adopt these technologies.

Conclusion

In conclusion, microbial solutions hold great promise for the future of sustainable agriculture. The use of microbial inoculants, biofertilizers, and biocontrol agents can improve soil health and plant growth while reducing the need for synthetic fertilizers and pesticides. Plant growth-promoting rhizobacteria, mycorrhizal fungi, and genetically engineered microbes are some of the most promising microbial solutions for sustainable agriculture.

However, there are also challenges to the widespread adoption of microbial solutions. The lack of standardized testing and regulation of microbial products can lead to uncertainty about their effectiveness and safety. Additionally, more research is needed to understand the interactions between microbial communities and their host plants, and to develop microbial solutions that are tailored to specific crops and soil conditions.

To fully realize the potential of microbial solutions for sustainable agriculture, collaboration between scientists, farmers, and policymakers is essential. Scientists must work closely with farmers to develop microbial solutions that are effective and practical. Policymakers need to create a regulatory environment that promotes the safe and responsible use of synthetic biology in agriculture. Finally, farmers need to be educated about the benefits and risks of microbial solutions and be given the tools and resources to adopt these technologies.

Overall, microbial solutions offer a promising alternative to conventional agricultural practices that can harm soil health and productivity. With continued research and development, microbial solutions have the potential to transform agriculture into a more sustainable and environmentally friendly industry.

References

- Bhattacharyya, P. N., Jha, D. K., Kumar, A., & Dubey, S. K. (2019). Microbial solutions for sustainable agriculture. Journal of pure and applied microbiology, 13(4), 2037-2047.
- FAO. (2018). The future of food and agriculture: Alternative pathways to 2050.
- FAO. (2019). The state of the world's biodiversity for food and agriculture
- IPCC. (2019). Climate Change and Land: An IPCC Special Report on climate change, desertification, land degradation, sustainable land management, food security, and greenhouse gas fluxes in terrestrial ecosystems.
- Forney, L. J., Zhou, X., & Brown, C. J. (2004). Molecular microbial ecology: land of the one-eyed king. Current opinion in microbiology, 7(3), 210-220.
- Rao, A. V., & Patra, D. D. (2009). The role of soil microbial diversity in determining soil fertility. Current Science, 96(2), 169-177.
- Torsvik, V., & Thingstad, T. F. (1996). Prokaryotic diversity--magnitude, dynamics, and controlling factors. Science, 296(5570), 1064-1066.
- Abdul-Baki, A. A., & Teasdale, J. R. (1993). Sustainable production systems for organic agriculture. Sustainable Agriculture Research and Education, 4(3), 38-45.
- Bardgett, R. D., & van der Putten, W. H. (2014). Belowground biodiversity and ecosystem functioning. Nature, 515(7528), 505-511.
- Jose, S. (2009). Agroforestry for ecosystem services and environmental benefits: An overview. Agroforestry Systems, 76(1), 1-10.

- Khan, S., Cao, Q., Zheng, Y. M., Huang, Y. Z., & Zhu, Y. G. (2017). Health risks of heavy metals in contaminated soils and food crops
- Banerjee, S., Kirkby, C.A., Schmutter, D., Bissett, A., Kirkegaard, J.A., and Richardson, A.E. (2016). Network analysis reveals functional redundancy and keystone taxa amongst bacterial and fungal communities during organic matter decomposition in an arable soil. Soil Biology and Biochemistry, 97, 188-198.
- Liang, Y., Cui, X., Sun, B., and Liu, X. (2017). Bacterial community diversity in the soil of a Chinese green pepper plantation using Illumina MiSeq sequencing: Effects of long-term fertilization and plant age. Scientific Reports, 7(1), 1-10.
- Compant, S., Duffy, B., Nowak, J., Clément, C., & Barka, E. A. (2005). Use of plant growth-promoting bacteria for biocontrol of plant diseases: principles, mechanisms of action, and future prospects. Applied and environmental microbiology, 71(9), 4951-4959.
- Kloepper, J. W., Ryu, C. M., & Zhang, S. (2004). Induced systemic resistance and promotion of plant growth by Bacillus spp. Phytopathology, 94(11), 1259-1266.
- Mohammadi, K., Sohrabi, Y., Pourreza, J., & Rezaei, K. (2017). The effect of biofertilizers on yield and yield components of wheat: a meta-analysis. Symbiosis, 73(3), 169-177.
- Raza, W., Zhang, R., Chen, X., Wang, J., Huang, Q., & Shen, Q. (2017). Evaluation of the plant growth-promoting activity of Pseudomonas putida UW4 and its interaction with wheat roots. Plant and Soil, 419(1-2), 27-45.
- Subler, S., Brandt, R., & Mullen, R. (2014). Vermicompost effects on yield, size, and quality of greenhouse tomatoes. HortScience, 49(3), 335-339.
- Vessey, J. K. (2003). Plant growth promoting rhizobacteria as biofertilizers. Plant and soil, 255(2), 571-586.
- Bashan, Y., & de-Bashan, L. E. (2010). How the plant growth-promoting bacterium Azospirillum promotes plant growth--a critical assessment. Advances in agronomy, 108, 77-136.

- Berg, G., & Smalla, K. (2009). Plant species and soil type cooperatively shape the structure and function of microbial communities in the rhizosphere. FEMS microbiology ecology, 68(1), 1-13.
- Bhattacharyya, P. N., & Jha, D. K. (2012). Plant growth-promoting rhizobacteria (PGPR): emergence in agriculture. World journal of microbiology and biotechnology, 28(4), 1327-1350.
- Gopalakrishnan, S., Srinivas, V., Alekhya, G., Prakash, B., Kudapa, H., & Varshney, R. K. (2015). Evaluation of Streptomyces strains isolated from herbal vermicompost for their plant growth-promotion traits in rice. Microbiological research, 173, 22-29.
- Hartman, K., van der Heijden, M. G., Wittwer, R. A., Banerjee, S., & Walser, J. C. (2018). Schistosoma mansoni infection reduces efficacy of multiple anthelmintic drugs in infected mice. Parasitology research, 117(9), 2947-2954.
- Jangid, K., Williams, M. A., Franzluebbers, A. J., Schmidt, T. M., Coleman, D. C., & Whitman, W. B. (2008). Land-use history has a stronger impact on soil microbial community composition than aboveground vegetation and soil properties. Soil Biology and Biochemistry, 40(6), 1457-1464.
- Kumar, A., Sharma, S., Mishra, S., & Kumar, V. (2016). Bacillus thuringiensis (Bt) transgenic crop: An environment friendly insect-pest management strategy. Journal of Environmental Biology, 37(2), 225-232.
- Philippot, L., Raaijmakers, J. M., Lemanceau, P., & van der Putten, W. H. (2013). Going back to the roots: the microbial ecology of the rhizosphere. Nature Reviews Microbiology, 11(11), 789-799.
- Richardson, A. E., Barea, J. M., McNeill, A. M., & Prigent-Combaret, C. (2009). Acquisition of phosphorus and nitrogen in the rhizosphere and plant growth promotion by microorganisms. Plant and Soil, 321(1-2), 305-339.
- Subler, S., Kettler, T. A., & Cowan, J. E. (2014). The benefits of using organic amendments. University of Illinois Extension.

- Vessey, J. K. (2003). Plant growth promoting rhizobacteria as biofertilizers. Plant and soil, 255(2), 571-586.
- Verma, J. P., Jaiswal, D. K., & Kumar, A. (2019). Microbial inoculants in sustainable agricultural productivity: Vol. 2: Functional applications. Springer.
- Huang, J., Chen, B., & Chen, C. (2021). Advances in formulation technology for microbial inoculants. Frontiers in Microbiology, 12, 702850.
- Kumar, A., Patel, J. S., & Bahuguna, R. N. (2020). Plant growth-promoting rhizobacteria: recent advancements and future challenges. Journal of Plant Growth Regulation, 39(4), 1189-1209.
- Tiwari, S., Singh, D., Singh, R., & Singh, M. (2021). Improved formulations of microbial inoculants for crop productivity: a review. Environmental Sustainability, 4(1), 1-17.
- Zhang, J., Wang, T., Liang, Y., Cheng, J., Chen, X., & Zhang, Y. (2018). Isolation and identification of plant growth-promoting rhizobacteria from wheat rhizosphere and their effect on wheat growth promotion. Microorganisms, 6(3), 69.
- Ahmad, F., Ahmad, I., & Khan, M. S. (2011). Screening of free-living rhizospheric bacteria for their multiple plant growth promoting activities. Microbial Research, 3(1), 1-7.
- Bashan, Y., & de-Bashan, L. E. (2010). How the plant growth-promoting bacterium Azospirillum promotes plant growth - a critical assessment. Advances in Agronomy, 108, 77-136.
- Berg, G., & Smalla, K. (2009). Plant species and soil type cooperatively shape the structure and function of microbial communities in the rhizosphere. FEMS Microbiology Ecology, 68(1), 1-13.
- Gopalakrishnan, S., Srinivas, V., Alekhya, G., Prakash, B., Kudapa, H., & Varshney, R. K. (2015). Evaluation of Streptomyces strains isolated from herbal vermicompost for their plant growth-promotion traits in rice. Microbial Cell Factories, 14(1), 1-10.
- Köhl, J., Kolnaar, R., Ravensberg, W. J., & Termorshuizen, A. J. (2019). Reduced risk management of plant pests and diseases in

organic farming: opportunities and limitations. Sustainability, 11(10), 2852.

- Philippot, L., Raaijmakers, J. M., Lemanceau, P., & van der Putten, W. H. (2013). Going back to the roots: the microbial ecology of the rhizosphere. Nature Reviews Microbiology, 11(11), 789-799.
- Smith, S. E., & Read, D. J. (2008). Mycorrhizal symbiosis. Academic Press.
- Verma, S. K., Kumar, R., Kumar, P., Tripathi, R. P., & Rani, R. (2019). Biocontrol agents and their secondary metabolites in the suppression of plant pathogens: a review. Biological Control, 144, 204-220.
- Glick, B. R. (2012). Plant growth-promoting bacteria: mechanisms and applications. Scientifica, 2012.
- Kumar, V., Yadav, S. K., Singh, A., & Kumar, V. (2019). Microbial enzymes in soil: potential strategies for improving soil fertility and crop productivity. In Microbial Enzymes in Bioconversions of Biomass (pp. 247-268). Springer, Singapore.
- Li, J., Sun, Q., Li, S., Zhang, J., & Hu, X. (2017). Engineering microbial consortia for high salt tolerance. Applied microbiology and biotechnology, 101(21), 7829-7839.
- Makarova, K. S., Zhang, F., & Koonin, E. V. (2020). CRISPR-Cas, the ubiquitous adaptive immune systems of bacteria and archaea. Journal of biological chemistry, 295(49), 17148-17167.
- The Royal Society. (2018). Realising the potential of synthetic biology: the Royal Society's perspective. London: The Royal Society.

Amity Institute of Microbial Technology
Amity University, Noida (U.P)
email : ishanggc@gmail.com

14. Sustainable Urbanisation and Making of Smart & Ressilient Green City : Case Study of New Town, Kolkata in Context of Recent Lulc Change at A Geospatial Glance

Sudip Dey

Abstract

Rajarhat New Town Kolkata is locality in the North 24 Parganas district. It is located in 88.43E to 88.60E and 22.64N to 22.50N, covering an area of 34.97 km² . It is close to Kolkata and also part of the area covered by the New Town Kolkata Development Authority (NKDA). Kolkata New Town is emerging in agricultural area through filling of big water bodies and converting cultivated lands into built up area. New Town Kolkata a recently developed major planned satellite township located in the peri-urban Kolkata & along its eastern outskirt. With this urbanization trend New Town has experienced many socio-economic changes. This study aims to analyse the change of LULC , Geomorphology ,Climatic Properties, Sustainable Urban Sprawl Expansion Of Town with help of Geospatial Techniques are Transformation and encounter different challenges. Such changes include literacy rate, availability of paved and metal road, electricity facility etc. The present study suggests to consider the possible micro-climatic changes in town planning for the sustainable development.

Keywords : Lulc, Gis , Urban Planning, Artificial Intelligence , Smart City , Suburbs , Cbd

Introduction

In the context of crowding, congestion in Kolkata metro CBD, Department of Housing, Govt. of West Bengal constituted a Technical Committee in May 1993 for preparation of a preliminary report of the New Town planning accordingly. In 1994 CMDA prepared a concept Plan for development of Kolkata New Town at the East of Kolkata Metro city and adjacent to the East Calcutta Wetland. The master plan was made by Indian Institute of

Technology in 1996. In 1999 the Master Land use Plan and detail Sector Plan for township was prepared and West Bengal Housing Infrastructure Development Corporation Limited (WB-HIDCO) formed as plan implementing body. Initially 3075 hectares of land has been acquired by the Department of Land and Land Reform (North and South 24 Parganas, West Bengal of which 68% was agricultural land. The area which has been chosen for development of Kolkata New Town is confined in between East Kolkata Wet Land in South-West and the Bidyadhari River in the east and hence is morphologically sensitive to natural flowage of the region. The findings of the study reveal that the planned urban growth has not really taken into account the high structural risk, probable obstruction of natural runoff, and has gone for an extensive horizontal and vertical expansion along with gradual clustering and compaction. Lastly, the chapter attempts to chalk out feasible policy interventions to maintain sustainable urbanization.

New Town, Rajarhat, a satellite township touted as the first "green, eco-friendly, self-sufficient, and smart city" of the state of West Bengal, was initiated in 1993 to ease the population burden of Kolkata megacity, both in terms of residential and commercial. This chapter mainly tries to assess the magnitude of anthropogenic interventions through rampant urbanization in New Town, Rajarhat, by theorization of urban expansion; contextualization of urban genic activities; identification of physio-structural issues and analysis of increasing urbanizing trend; delineation of spatio-temporal change of land use and land cover (LULC) over a span of almost 40 years; and identification of urban clusters with prediction of future trajectory. The methodology involves literature review, 3-D mapping along with rapid visual screening (RVS) survey, LULC mapping and change detection (1980–2020), Shannon's entropy analysis, Normalized Difference Built-up Index (NDBI), hotspot analysis, identification of urban patches, and anticipation of trends.

Aim & Objective

The trend of establishment of New Town has been fortified by Smart City Mission, which started on 25th June, 2015 to promote sustainable and inclusive cities which will provide core infrastructure along with clean and sustainable environment for decent quality of life. Here an attempt has been taken to focus on the land use change because of implementation of planned New Town in Kolkata and its associated effects on environment as well as on the people living there has been deliberated.

- To identify the urban growth of the study area during last two decades.
- To analyse the Dynamic change & shift of Industries from CBD To Urban Fringe with a context of LULC change detection.
- To analysis the socio-economic structure of the New Town.

Literature Review

Joy Karmakar (2015) Planning of the peri-urban area is one of the significant challenges in coming decades that most of third world cities have to face. Rajarhat New Town a recently developed major planned satellite township located in the peri urban areas of Kolkata. The paper have two main sections first part dealt with the evolution of the periurban areas of Kolkata and second part composed of the brief history of Rajarhat area, land acquisition process and the role of the state, urbanization process and its socio-economic implications of the Rajarhat with special emphasis on environment destruction.

Amar Biswas, Dr. Omveer Singh (May 2017) , It consists of the two erstwhile villages Rajarhat and Bhangar (located in South 24 Parganas), which is now statutory planned development area.The present study is divided into two parts. The first part is non analytical covering the genesis of New Town. The second part covering analytical part whereby many local aspects has been studied intensively. Due to planned urbanization in the study area, there is a significant change in land use land cover and as well as in demographic structure.

Dr. Abhijit Pandit (May-june 2018) Encroachment of human settlement with the expansion of city area became the major concern for the environment and climate. The expansion by sprawl in planned pattern occupied the north-eastern fringe of Kolkata by development of New Town area. The entire development impacted the rise of temperature along with impervious surface area which could be considered as the prime parameters in analysing the Urbanization and present day environmental condition. The study depicts the analysis of Urban Heat Island condition and its predicts probable impact on future climatic condition of Newtown location in Kolkata

Location of Study Area :

Location of the town within Bidyadhari Basin area is mapped (Fig-2) by superimposing the base map of New Town (HIDCO) on satellite image (Landsat 8, OLI). Dum Dum, Salt lake City of Kolkata and East Kolkata Wetland are in the west of the Kolkata New Town. Bhangar, Polerhat municipalities of Bhangar Block of South 24 Parganas District are located in the east and Nowi River marks the north-east boundary of the town (Fig-1). The town is located within the Bidyadhari Basin area. Its position within the basin area is shown in fig 2. It is demarcated by HIDCO. Total area is 102 sq. km from which 60 sq. km is now under planning. Altogether twenty four mouzas from two blocks (Rajarhat of North 24 Parganas and Bhangar-II of South 24 Parganas of west Bengal) were included for this project. The project area covers 6035 hectors area of land in 2015. The new city is a planned project of the Govt. of West Bengal. The New Town Development Act came into effect in 2007.

Figure : 1 Location Mapping Layout of New Town Kolkata

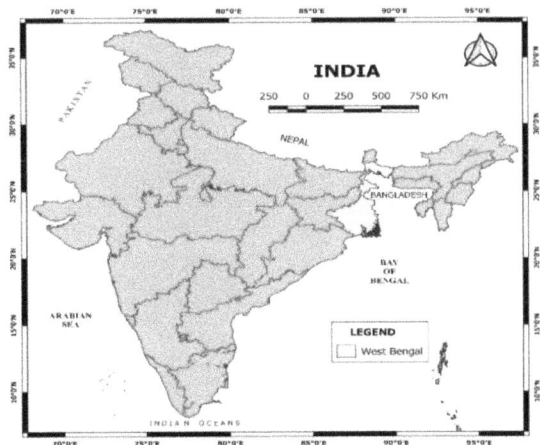

Sustainable Consumption and Production

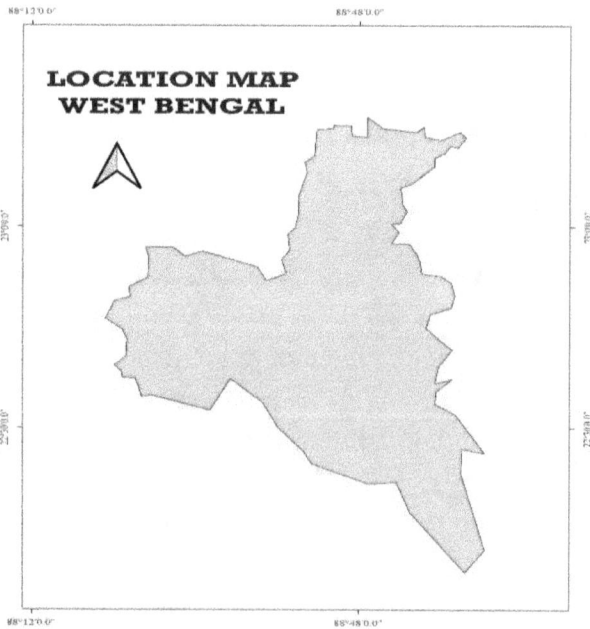

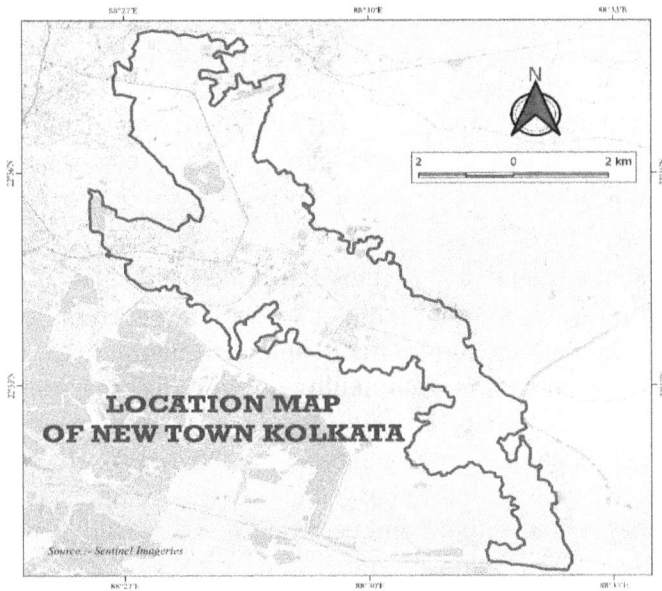

Geographical History of Rajarhat New Town:

The case of New Town is the first planned production of suburbs in the post liberalising context of Bengal. It is a complex area of study with a whole raft of, often divergent, issues interacting at many levels. New Town presents economies of scale that make it larger than life, covering an area three times larger than Salt Lake and an area covering 8773 acres for up to 5 million people .

In a strange reversal of responsibility, the genesis of New Town lay with KMDA's Concept Plan 1993; however, its implementation is undertaken by the state government, which signifies the perceived importance of the project regionally and that stakes are high. In 1999 the state government established a parastatal body called WBHIDCO for a 'shrewd' execution of New Town, in that WBHIDCO sits at 'arm's length' to the state government—for the necessary manoeuvring and political legitimising of the state government's actions. WBHIDCO has secured the entire cost of New Town development from the capital market (11) (Biswas, 2006), making it one of the largest off-budget schemes(12) in the country in recent years.

It has powers to develop the entire range of infrastructure services, construct housing and commercial premises, and acquire and sell land for different purposes. The first phase of the project—Action Area 1, which covers 6655 acres of land—is nearing completion. In the following paragraphs I delve deeper into the issues concerning land use and regulations, state–market alliance in acquisition, and approach to affordability, to take the conversation on land development further.

Physical Aspects of Study Area

❖ Ecological Concerns over Township Creation

This section throws a light on the changing land uses and ecological consequences encountered by the people of Rajarhat and New Town in particular. Conversion of agricultural land into urban land did not only bring economic change but also brought paramount environmental change. Environmental change includes the vanishing of agriculture and wetland ecosystems as a substantial part of the mouza were under wetland and ecosystem. Ecosystem services that the villagers used are at stake now. Not only has the acquired land been transformed, but the area which is not acquired is also transformed for the development of housing enclaves.

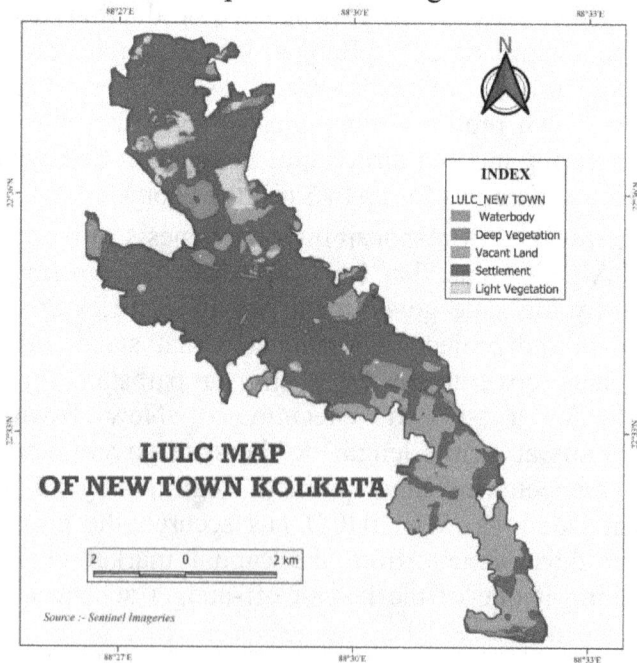

Figure 2 Lulc Map of New Town,Kolkata

136

Dhar et.al. (2019) studied the land surface temperature change due to change in the landuse in Rajarhat block under which New Town Kolkata and surrounding villages are located. They find out 2 that from 1990 to 2016: 13 km of vegetation cover 2 lost due to urbanization; 9.3 km of open land converted to agricultural land and open 2 fields/parks; 1.4 km of aquaculture ponds 2 converted to tree cover/scrublands and 1.45 km of lakes/ponds filled up. Furthermore, due to this change in land-use pattern over 26 years, Land 0 Surface Temperature has increased by 0.94 C. The urban-heat-island (UHI) phenomenon has also increased.

❖ **Lulc Change Detection of New Town**

The following Map 1 shows that there are still a few patches of agricultural land that exists outside the smart city New Town Kolkata. Likewise, only a few artificially created large water bodies exist within the township.

Table 1:- LULC Classification of New Town

Class	PixelSum	Percentage %	Area [metre^2]
1 Unidentified	19668	5.148166	1966800
2 Waterbody	2989	0.782381	298900
4 Deep Vegetation	126	0.032981	12600
5 Vacant Land	76581	20.04534	7658100
7 Settlement	265621	69.52719	26562100
11 Light Vegetation	17054	4.463942	1705400

LULC COVERAGE OF NEW
TOWN,KOLKATA 2022

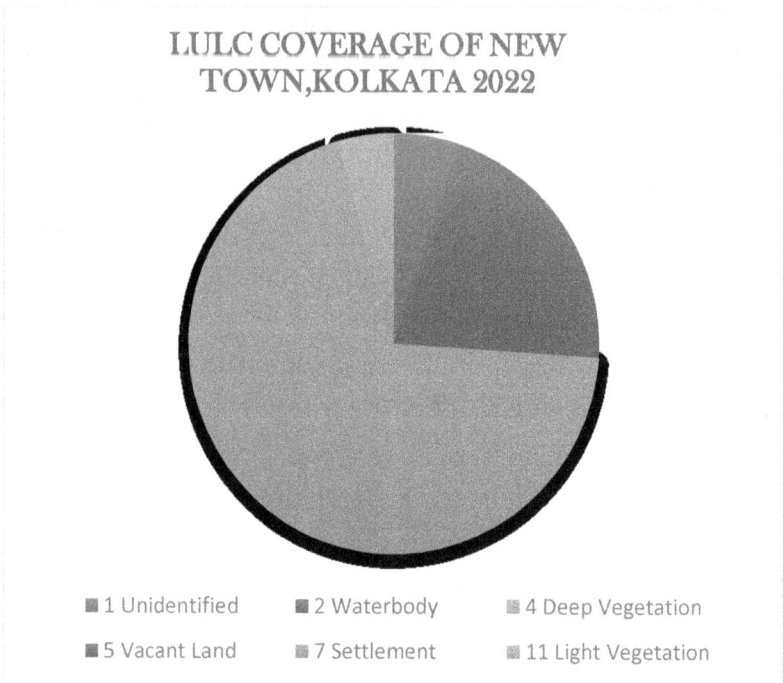

■ 1 Unidentified   ■ 2 Waterbody   ▓ 4 Deep Vegetation

■ 5 Vacant Land   ▓ 7 Settlement   ▓ 11 Light Vegetation

Figure 3 : LULC Areal Coverage Percentage In New Town,
Kolkata

It is to be noted that most of the converted agricultural land within
and outside the juridical boundary of New Town Kolkata are now
transformed into several housing estate like Sanjeeva Orchard,
Sukhobrishti Township, Unitech Uniworld, etc. For example, in
Thakdari 2 mouza, apart from the East Kolkata Wetland (EKW) area
(35.60 acre approximately) within the village, mouza had 1.62 acre
wetland area and two big Beels are there, whose size is 0.69 acre
and 0.11 acre. Out of 1.62 acres of wetland in the village, 0.7 acres
of land has been directly purchased by WBHIDCO for the New
Township Project (Government of West Bengal, 2013). In total ten
ponds and two small water bodies compare to a pond i.e. doba has
been filled up for purposes (Karmakar, 2015). However, for the
permission of any project, the state government created a legal body
called State Level Expert Appraisal Committee (SEAC) and they
look at the various aspects of the project including the
environmental aspects.

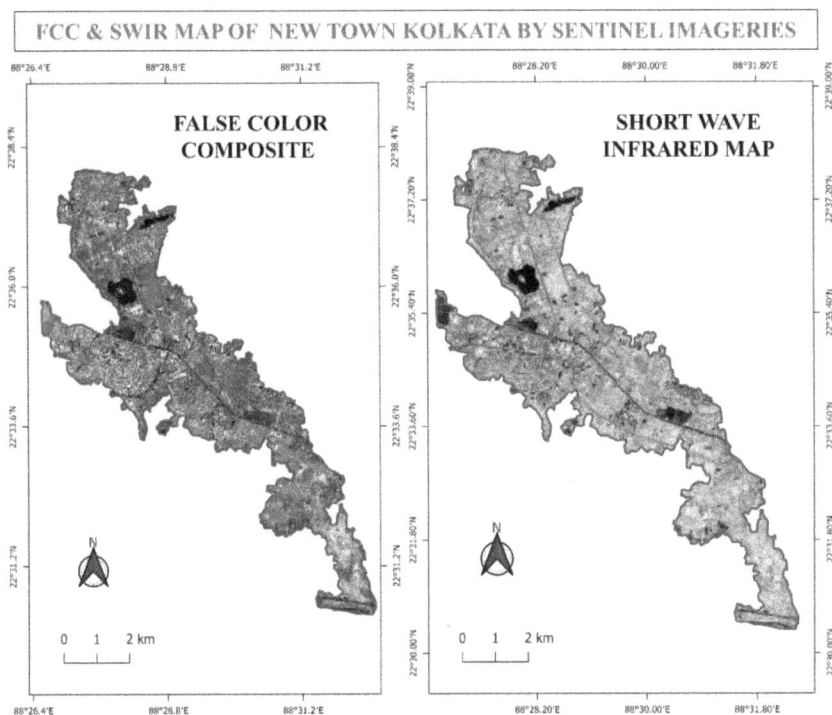

Figure 4 FCC & SWIR Map of New Town,Kolkata

The Committee proposes some stipulated conditions for environmental clearance as per the provision of Environmental Impact Assessment Notification 2006 and the subsequent amendments like water bodies, if any, should not be lined and their embankments should not be cemented. The water bodies are to be kept in natural conditions without disturbing the ecological habitat. No existing water body, if any, should be encroached/relocated/reshaped without prior permission of competent authorities and the unit should strictly abide by The West Bengal Trees (Protection and Conservation in Non Forest Areas) Rules, 2007. The proponent should undertake plantation of trees over at least 20% of the total area. No trees can be felled without prior permission from the Tree Cutting Authority constituted as per the West Bengal Trees (Protection and Conservation in Non Forest Areas) Act, 2006 and subsequent rules (GoWB, 2012). It is worthwhile to mention that these laws are applicable within the boundary of New Township, but outside the boundary, especially in peripheral villages, various housing enclaves are formed ignoring such environmental concerns.

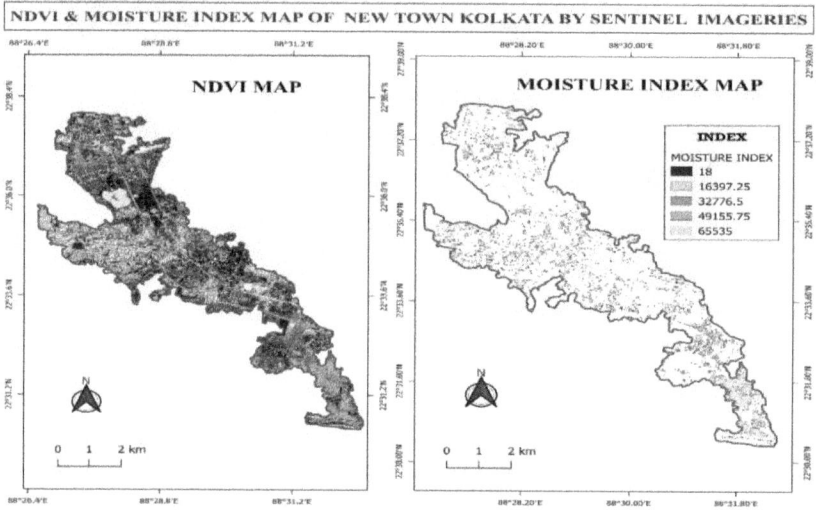

Figure 5 NDVI & Moisture Index Map 0f New Town Kolkata

It would be a future ready global services hub attracting the best talent with a fine work-life balance. Walkability and transport are the main areas of intervention, followed by safety & security, the economy and employment and water. The lack of adequate job opportunities for the residents or future property owners of New Town Kolkata is one of the main issues highlighted by the citizens.

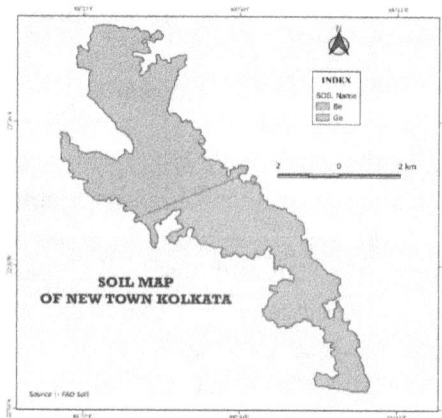

Figure 6 :- Soil &NDWI Map of New Town Kolkata

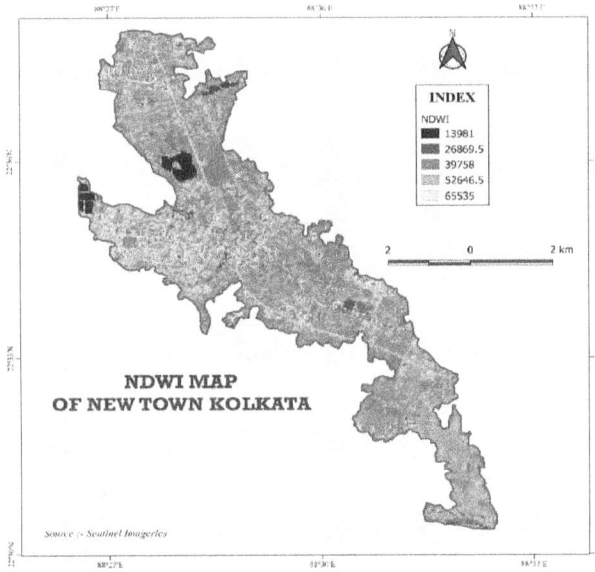

NDWI MAP
OF NEW TOWN KOLKATA

❖ Hydro-Geomorphology

Based on Water level data from *indiawris.gov.in portal Radar chart & Diagram* has been prepared to show the dynamic change of surgace water level of Rajarhat New town Kolkata,2022

Water Level from 2015 to 2019 of Rajarhat New Town

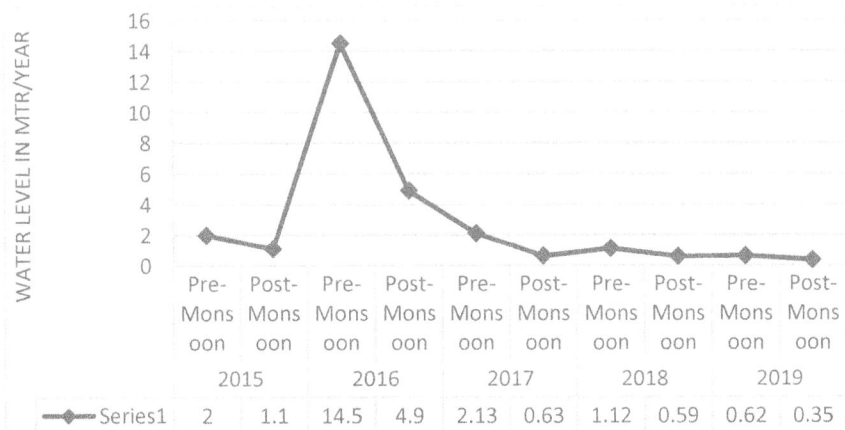

	Pre-Monsoon 2015	Post-Monsoon 2015	Pre-Monsoon 2016	Post-Monsoon 2016	Pre-Monsoon 2017	Post-Monsoon 2017	Pre-Monsoon 2018	Post-Monsoon 2018	Pre-Monsoon 2019	Post-Monsoon 2019
Series1	2	1.1	14.5	4.9	2.13	0.63	1.12	0.59	0.62	0.35

Source :- https://indiawris.gov.in/wris/

141

Fig : - 8 Radar Chart Showing Trend of Water Level from 2015 to 2019 of Rajarhat New Town Kolkata

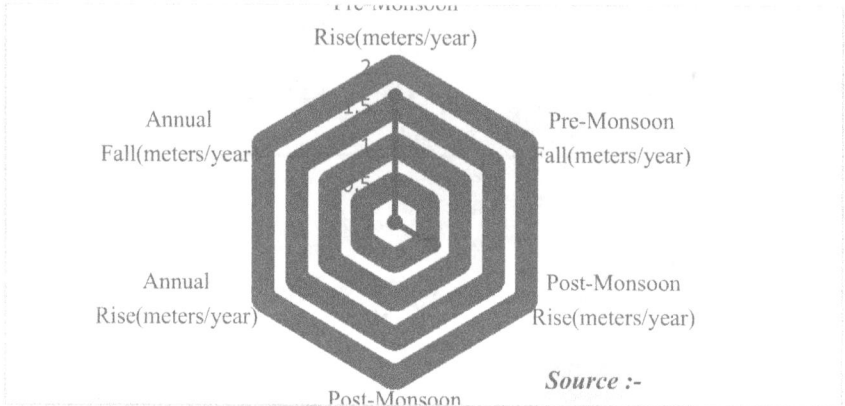

Urbanization is taking place rapidly in most parts of the world and urban population is growing at accelerated rate. Presently 54.4 per cent of world's population lives in urban areas and this is expected to rise to 60 per cent by 2030 and among top 31 world's mega cities Delhi ranks 2nd, Mumbai ranks 4th and Kolkata 14th (DESA 2016). India is the second largest populous country in the world. Though the level of urbanization is comparatively low (31.16 per cent), India accounts for 377.16 million urban people which is the second largest urban populated country in the world (MHUPA 2016).

Figure 9:- Geomorphology Map of New Town , Kolkata

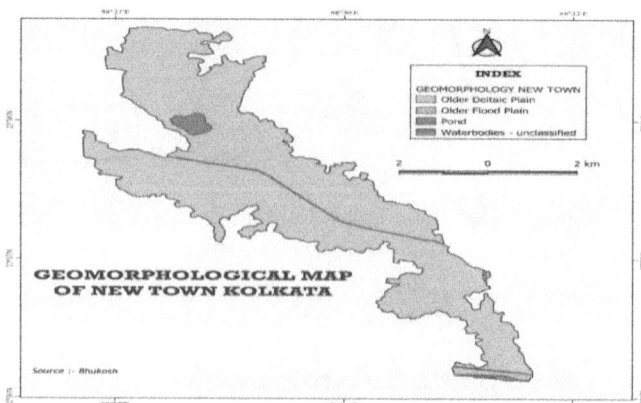

Economic Setup of The Area

Development project of such magnitude always attributes a considerable change in the socio economic pattern of livelihood when a truly agrarian society is forced to adopt a semi urban character. However to avoid disturbance to the residential areas and to keep the surrounding as well as intervening villages intact the planning and layout of the project has been made accordingly and it is projected that about 2500 persons would be displaced please wise . As such to pacify the social and political tourn oil, the state government has worked out a suitable compensation package on the basis of living standard of those displaced person. The compensation package has been calculated in terms of cost of food stuff. The total net amount earned out of cultivation of 759 acres in a year for non paddy crops and vegetables has been estimated to be at Rs. 34.92 lakh cabbage accounted for Rs. 14.63 lakh. As such after conversion of agriculture land for unlena usage , the gross cropped are being 11,351 acres out of about 6032 acres, the net revenue loss would be a handsome figure of about 90.0 lakh a year.

Figure 10 : GIS Centric City Management in New Town, Kolkata

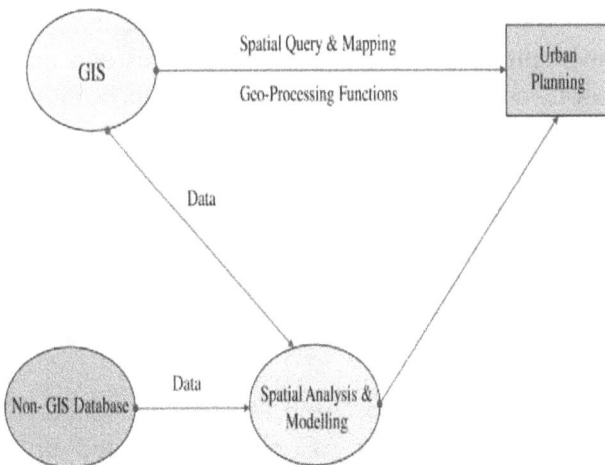

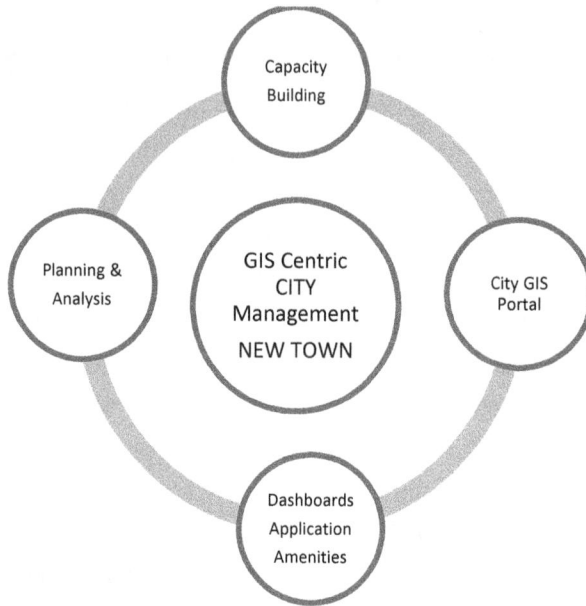

Capacity Building

Planning & Analysis

GIS Centric CITY Management NEW TOWN

City GIS Portal

Dashboards Application Amenities

Social Life of Resident

The residential areas will consist of group of residential cluster(i e group housing and plots) in which different sections of people of different income levels can live harmoniously without being detached from the whole community. Neighborhood units will be used as planning modules of residential growth but their sizes and forms would vary depending on various considerable factors. A neighborhood units based around some services and amenities like park , school, convenient shopping etc will be planned for a community of 2000 to 3000 people. Group of 3 to 5 such communities may be served by higher education and health facilities. There will be provision for services villages at different part of the New Town. The service village will provide facilities for work and living for the services population. Planning of the service village and it's dwelling units deserve special attention.

Connectivity & Accessiblity

The main mode of transportation, like other parts of Kolkata includes air conditioned, non AC government and private buses, Taxi and other popular India transport like autorickshaw, however the buses are sporadic and the routes originating out are very few. The inhabitants of Rajarhat Gopalpur enjoy multiple ways of transportation such as railways, roadways, port and air. The railway in the city comes under the division of Eastern Railways from Sealdah. The area is well connected by road ways. It is very close to National Highway 34 which connects South Bengal with North India. Major connecting roads are Eastern Metropolitan Bypass Road (EMB Road), Kaji Najrul Islam Sarani (KNIS) or VIP Road, Kona Expressway and Belgharia Expressway. Road Network is the basic transportation structure within NTK, although metro work is going on in full swing and in near future it will become important mode for communication.The nearest railway station to Rajarhat Gopalpur is Dum Dum Junction Railway Station, which is well connected to major cities of West Bengal. Besides that, Dum Dum Metro Station is also located near to this city with a distance of 5 km.

The city is placed aside a famous road i.e. Najrul Sarani previously known as VIP Road. One has to cover 11 km distance by coming via this road from Kolkata to Rajarhat Gopalpur. The city also provides benefits to the voyagers who like to come through port as Kolkata Port at Kidder pore is the nearest port to this town with a distance of 25 km. Moreover, the nearest airport to the city of Rajarhat Gopalpur is Netaji Subhas Chandra International Airport, located at a distance of 5 km. A graph has been drawn to represent the common means of travel availed by the people of the sample households .A moderate number of households ow bicycle and therefore that is one of the most common modes of transport. But the most frequent and common mode of transport is bus. Buses are easily accessed by the residents. E rickshaw are also easily availed by the most of the households and of course walking is one of the mode preferred by most of the households. New Garia Airport line of Kolkata Metro is supposed to pass through New Town by 2014. Nowadays E Cabs are very facilitated across New Town. Apart from well-connected transportation routes the city, the main arterial road is considered as Wi-Fi corridor as the road is Wi-Fi enabled

(Chakraborti, 2014). It stretches about 10.5 km from airport to Sector-V of Salt Lake City.

Major Findings

- The process of urbanization has taken place as what it was to what is now, has major difference. A town which was wet land now it is IT sector.
- The execution of well planned township was initiated by land acquisition from farmers & locals. As well to maintaining Environmental Balance & Urban Ecology Conservation Sustainable Green Building Must be Constructed.
- Initially steps were hard but after effects are improving the daily life and socio economic environment of Gopalpur locals. Smart Infrastructure has been well distributed.
- Surveyed area lacks in sewage and potable water. Transport Accessibility is Good. West Bengal Silicon Valley will grow up in here. It will be Upcoming Major Industrial IT hub of Eastern India.
- Government should plan something new for seasonal water logging as it effects the locals badly & City Has Achieved Many Pledge Like Cycle City ,Green City.

Limitation of The Study

- Due to the cooperative apartments and society interaction with middle upper class is limited. Newly developed residential areas, which are in maximum cases here apartments or complex buildings, respondents of these areas were reserved and did not provide full cooperation.
- The study area has repeatedly undergone changes in its boundary; there is a repetition of inclusion and exclusion of area within it which creates difficulty during the preparation of maps
- Local people are not so vocal for the political significance limits the data
- The present study is totally present by hybrid Methodology , data collected from different Secondary Sources.

Meanwhile, in the present, most corporate employees leave the area as soon as their working hours are done, while local residents are split between those who live in the slums and make a living in the informal economy, and those who segregate themselves between gated communities and shopping mall

Conclusion :

City planning is often done in normative fashion with minimum concerns of local environmental issues. The Kolkata New Town is a well-planned township. The sectoral plan as Central Business District, industrial area, residential area, road network and other aspects have been well comprehended for the Master Planning. It can be said that decades ago this area was a vacant land, part of East Kolkata wet land and agriculture land of North 24 parganas,. Over a period of time it has gone through various phases of development. The study area is under process to form modern satellite city of west Bengal, India. The land use land cover 2022 depicts the maximum utilization of land for settlement due to rapid growth of population over a period of time. Agriculture land has been cut off massively to construct multi storage building , complexes, shopping malls etc. But it is evident from this study that the plan is mindless about micro morphometry of the area. As a result, this emerging township is impeding the natural flowage of the area and is exaggerating waterlogging problem of surrounding region.

References

Chakrabarty, Kakali et.al, 2015,Land People And Power, Gyan Publishing House .

Desai, Manisha and Patel, J. N. 2014, Use of Digital Elevation Model to compute Storm Water

Dey, I. Samaddar, R. and Sen, S. K. "Beyond Kolkata: Rajarhat and the Dystopia of the Urban Imagination" Routledge (New Delhi) (2013)

Karmakar, J. (2015) Encountering the realityof the planning process in peri urban areas of Kolkata: Case study of Rajarhat, Archives of Applied Science Research, 7(15

Kundu, R. "Making Sense of Place in Rajarhat New Town the Village in the Urban and the Urban in the Village". Economic & Political Weekly, LI (17). 93-101(2016).

Mallik, C. "Land Dispossession and Rural Transformation: The Case of Fringe Villages of Kolkata". Journal of Rural Development. 33 (1). 51- 71(2014).

Roy, A. "Development, Land Acquisition and Changing Facets Of Rural Livelihood: A Case Study From West Bengal". Journal of Rural Development. 33(1), 15-32. (2014).

Sen, Dhrubajyoti. (2013) Real-time rainfall monitoring and flood inundation forecasting for the city of Kolkata, ISH Jour. Hydraul. Engg., v.19(2), pp.137–144.

Sengupta, U., & Sharma, S. (2007). Government intervention and land development in New Town, Kolkata: Emerging Lessons for the Policy Makers? Retrieved 2017, from http://siteresources. worldbank.org/Inturbandevelopment/Resources/336387-1269364687916/6892589-1269394475210/ sharma.pdf

Shao, Z. (2015). The New Urban Area Development. New York: Springer. "pp. 3-21".

Taylor, G. R. (1915). Satellite Cities: A Study of Industrial Suburbs. New York and London: D. Appleton and Company. "pp. 26""

West Bengal Housing and Infrastructure Development Corporation. New Town Calcutta, Project Report. Kolkata, India. (1999).

Xiao, J., Y. Shen, J. Ge, R. Tateishi, C. Tang, Y. Liang, and Z. Huang. (2006). Evaluating urban expansion and land use change in Shijiazhuang, China, by using GIS and remote sensing. Landscape and Urban Planning (Elsevier) 75, 69–80.

Yan, H., J. Liu, H. Q. Huang, and M. Cao. (2009). Assessing the consequence of land use change on agricultural productivity in China. Global and Planetary Change (Elsevier) 67, 13-19.

**Msc. Student In Geography ,
Asutosh College**

15. Sustainable Development in India : An Overview

Dr.Yogendra Kumar

Abstract

In this chapter sustainable development highlights the concept of economic growth that meets the needs of the present generation without compromising the ability of future generations to meet their own needs. The goal of sustainable development is to balance economic, social, and environmental objectives to promote development that is socially, economically, and environmentally sustainable. India has launched several initiatives to promote sustainable development, but there are still challenges that need to be addressed, such as environmental degradation and the need to balance economic growth with environmental sustainability. Experts in the field have defined sustainable development as development that meets the needs of the present without compromising the ability of future generations to meet their own needs, while ensuring that economic, social, and environmental factors are taken into account. Achieving sustainable development requires the cooperation and involvement of all sectors of society, including government, businesses, civil society, and individuals.

Introduction

Sustainable development in India is an important topic given the country's size, population, and economic growth. India has made significant progress towards sustainable development in recent years, but there are still many challenges that need to be addressed. One of the main challenges facing India is the need to balance economic growth with environmental sustainability. While India has made progress in reducing poverty and increasing economic growth, this has often come at the cost of environmental degradation, such as air and water pollution, deforestation, and soil degradation. Addressing these environmental issues is important for the long-term sustainability of the country.

To achieve sustainable development, India has launched several initiatives such as the Swachh Bharat Abhiyan (Clean India Mission), Smart Cities Mission, and National Solar Mission, among others. These initiatives aim to promote sustainable development by addressing issues related to sanitation, urban development, and renewable energy.

Additionally, India has made efforts to reduce its carbon footprint through the adoption of clean energy sources, such as wind and solar power, and has pledged to reduce its greenhouse gas emissions intensity by 33-35% by 2030, compared to 2005 levels, as part of its commitment to the Paris Agreement. Overall, sustainable development is a crucial aspect for India's future growth, and the country has taken several steps to promote it. However, there is still much work to be done to address environmental challenges and ensure a sustainable future for all citizens.

Meaning and Definition of Sustainable Development :

Sustainable development refers to a type of economic growth that can meet the needs of the present generation without compromising the ability of future generations to meet their own needs. In other words, sustainable development seeks to balance social, economic, and environmental objectives, so that economic growth can be achieved without harming the environment or depleting natural resources. Sustainable development is often defined as development that meets the needs of the present without compromising the ability of future generations to meet their own needs. It is a holistic approach that takes into account the economic, social, and environmental dimensions of development.

The concept of sustainable development was first introduced in the 1987 report "Our Common Future" by the World Commission on Environment and Development. The report defined sustainable development as "development that meets the needs of the present without compromising the ability of future generations to meet their own needs." Sustainable development is a long-term approach that recognizes the interconnectedness of social, economic, and environmental systems, and seeks to promote development that is socially, economically, and environmentally sustainable. It involves balancing economic growth with environmental sustainability and social equity, so that present and future generations can enjoy a high quality of life.

Sustainable development is based on the idea of intergenerational equity, which means that the needs of the present generation must be balanced with the needs of future generations. It recognizes that economic development and environmental protection are not mutually exclusive, and that environmental sustainability is crucial for economic growth in the long term.

Sustainable development involves a broad range of activities, such as conserving natural resources, reducing greenhouse gas emissions, promoting renewable energy, improving access to education and healthcare, reducing poverty, promoting gender equality, and ensuring social inclusion. It is an approach that requires the cooperation and involvement of all sectors of society, including government, businesses, civil society, and individuals. Sustainable development is also closely linked to the concept of sustainable consumption and production, which involves reducing the environmental impact of production and consumption patterns, while promoting sustainable lifestyles and resource use. This involves the adoption of cleaner and more efficient production methods, the reduction of waste and pollution, and the promotion of sustainable consumption patterns.

Here are some definitions of sustainable development:

The **United Nations' Brundtland Commission** (1987) defines sustainable development as "development that meets the needs of the present without compromising the ability of future generations to meet their own needs."

The **International Institute for Sustainable Development** (IISD) defines sustainable development as "development that meets the needs of the present without compromising the ability of future generations to meet their own needs, while ensuring that economic, social, and environmental factors are taken into account."

The **World Business Council** for Sustainable Development (WBCSD) defines sustainable development as "the continuing improvement of people's quality of life while preserving the integrity of the earth's ecological systems."

The **Global Footprint Network** defines sustainable development as "meeting the basic needs of all people and ensuring that all people have equal opportunity to achieve their full human potential, while living within the ecological limits of the planet."

The **United Nations Development Programme** (UNDP) defines sustainable development as "development that promotes economic growth, social well-being, and environmental protection and enhancement, while taking into account the needs of future generations."

These definitions all emphasize the importance of balancing economic, social, and environmental objectives, while ensuring that the needs of present and future generations are met. They recognize that sustainable development requires a long-term approach and the cooperation of all sectors of society.

The Indian government has taken several steps to achieve the Sustainable Development Goals (SDGs) by 2030. Some of these steps include:

1. **Implementation of National Action Plan on Climate Change (NAPCC)** : The NAPCC outlines eight missions to address climate change, including the National Solar Mission, the National Mission for Enhanced Energy Efficiency, and the National Mission on Sustainable Habitat.

2. **Launch of Swachh Bharat Abhiyan** : The Swachh Bharat Abhiyan, or Clean India Mission, aims to achieve a clean and open-defecation-free India by 2nd October 2019. It focuses on building toilets, solid waste management, and promoting cleanliness and hygiene.

3. **National Health Mission** : The National Health Mission aims to provide accessible, affordable, and quality health care to all citizens of India. It focuses on reducing maternal and child mortality rates, improving immunization coverage, and providing universal access to basic health services.

4. Launch of Smart Cities Mission : The Smart Cities Mission aims to develop 100 smart cities across the country that are sustainable, livable, and inclusive. It focuses on providing quality infrastructure, promoting sustainable urban development, and improving the quality of life of citizens.

5. Promotion of Renewable Energy : The Indian government has set an ambitious target of generating 450 GW of renewable energy by 2030, which includes 175 GW of renewable energy capacity by 2022. The government has also launched several programs and schemes to promote the use of renewable energy, including the National Solar Mission and the Green Energy Corridor Project.

6.Beti Bachao Beti Padhao Yojana : This program aims to improve the gender ratio by educating girls and promoting gender equality. It focuses on preventing female foeticide, promoting girls' education, and empowering women.

7. State Action Plans on Climate Change : The State Action Plans on Climate Change (SAPCC) aim to create institutional capacities and implement sectoral activities to address climate change. These plans are focused on adaptation with mitigation as co-benefit in sectors such as water, agriculture, tourism, forestry, transport, habitat and energy. So far, 28 states and 5 union territories (UTs) have submitted their SAPCCs to the MoEF&CC. Out of these, the SAPCCs of 32 states and UTs have been endorsed by the National Steering Committee on Climate Change (NSCCC) at the MoEF&CC.

These are just a few examples of the steps taken by the Indian government to achieve the SDGs. While there is still a long way to go, these initiatives demonstrate the country's commitment to sustainable development and provide a roadmap for achieving the SDGs by 2030Sustainable development in india

Conclusion

Achieving sustainable development requires the participation and commitment of individuals, communities, businesses, governments, and organizations at all levels. Here are some actions that individuals and communities can take to promote sustainable development:

Reduce, Reuse, and Recycle : One of the most effective ways to reduce waste and conserve resources is to reduce, reuse, and recycle. Individuals can reduce waste by choosing reusable products, repairing items instead of throwing them away, and recycling whenever possible.

Conserve Energy : Another way to promote sustainable development is to conserve energy. Individuals can do this by turning off lights and appliances when they are not in use, using energy-efficient appliances and light bulbs, and reducing unnecessary travel.

Support Sustainable Products and Services : Consumers can support sustainable products and services by choosing environmentally friendly products and supporting companies that have sustainable business practices.

Advocate for Sustainable Policies : Individuals can also advocate for sustainable policies at the local, national, and international levels. This includes supporting policies that promote renewable energy, reduce greenhouse gas emissions, protect natural resources, and support sustainable development.

Educate Yourself and Others : Finally, individuals can educate themselves and others about sustainable development and the importance of protecting the environment. This includes learning about the impacts of human activities on the environment and promoting sustainable lifestyles and behaviors.

These actions may seem small, but they can have a significant impact when implemented collectively. By taking these steps, individuals and communities can contribute to a more sustainable future for themselves and future generations.

Overall, sustainable development is a complex and multidimensional concept that requires a long-term approach and a commitment to balancing economic, social, and environmental objectives. It is an approach that seeks to promote development that is sustainable in the long term, and that can meet the needs of both present and future generations.

Reference

1. Osserwarde M.J. "Introduction to Sustainable Development" SAGE Publications Pvt. Ltd,2018
2. Jhingan M L "Environmental Economics : Theory, Management and Policy" Vrinda Publications 2009
3. https://www.researchgate.net/publication/358721142sustainable_ Development_In_India
4. https://www.jagranjosh.com/current-affairs/sustainable-development-and-india-1503408725
5. C. Surbha, "The benefits of sustainable development" Mar 2022

Assistant Professor
Govt. Girls'college Ajmer,
Rajasthan

16. Sustainability and Tourism

Dr. Achole Pandurang Bapurao

Abstract
There have been requests for a new conceptual framework for development as a result of concern over the detrimental consequences of development on the environment. This idea of "sustainable development" has emerged as the new paradigm for all types of development, including tourism.

The Term "Sustainable Development" was First Used
The Brundtland Report, formally the International Commission on Environment and Development report, is generally credited with popularising the term "sustainable development" (WCED, 1987). The term "sustainability," as opposed to "sustainable development," has its roots in conservation-related worries and can be linked to the conservation movement of the middle of the nineteenth century (Stabler and Goodall, 1996).The Global Conservation Strategy, issued in 1980 by the World Conservation Unit (IUCN), is where the term "sustainable development" first appeared (Reid, 1995)

Yet, the phrase did not become widely used until it was included in the Brundtland Report seven years later. This may be because by 1987, environmental consciousness had increased significantly. The Norwegian Prime Minister Gro Harlem Brundtland undertook an investigation into the state of the environment at the request of the United Nations General Assembly, which resulted in the report Sustainability and tourism • 149. Concerned about the impact of economic expansion since the 1950s on the environment, the United Nations commissioned a 22-person independent committee in 1984. people from a range of member countries, including both developed and developing nations, to develop long-term environmental plans for the global community (Elliott, 1994). The United Nations' main environmental concerns were the high rates of unsustainable resource consumption associated with development and the contribution of pollution to significant environmental issues like global warming and the ozone layer thinning, which harmed human well-being. The realisation that the environment and development

156

are inextricably linked came along with a greater awareness of environmental problems. A failing environment does not allow for development. resource base; similarly, the environment cannot be safeguarded if development disregards the consequences of environmental harm, as described in Chapter 4. Even though the word "development" is widely used today, academic study of the topic has only been practised since the 1950s, when colonial territories began to gain independence (ibid.). Although "growth" and "development" are frequently used interchangeably, there is a significant distinction between the two. Growth is the process of expanding or growing, whereas development is the improvement of a situation.

It was the predominance of the negative aspects of development changes that led to the calls for sustainable development. The term gained greater attention following the United Nations Conference on Environment and Development (UNCED), held in Rio de Janeiro in June 1992, popularly referred to as the 'Earth Summit'. At the Earth Summit a programme for promoting sustainable development throughout the world, known as Agenda 21, was adopted by participating countries. Agenda 21 is an action plan laying out the basic principles required to progress towards sustainability. It envisages national sustainable development strategies, involving local communities and people in a 'bottom–up' approach to development, rather than the 'top–down' approach which has typically characterised national development plans. Although tourism as an economic sector was not debated in Rio, five years later at the 'Earth Summit II' in New York, it was debated as a recognised economic sector. In the recommendations and outcomes of the report it was stated

Governments, international financial institutions, non-governmental organisations, the commercial sector, and academics all started using the term "sustainable development" in the last decade of the 20th century. When faced with tremendous ecological change brought on by human activity, sustainable development "has developed over three decades from an environmental issue to a socio-political movement for constructive social and economic change," as Farrell and Twining-Ward (2003: 275) note.

157

Sustainable Development Definition
The origins of sustainable development
- The term 'sustainable development' can be traced back to the conservation movements of the mid-nineteenth century
- Established as a policy consideration in the World Conservation Strategy published by the World Conservation Union (IUCN) in 1980
- Term gains greater attention and popularity after the publication of the Brundtland Report (1987)
- The 'Earth Summit' 1992 held in Rio de Janeiro adopts 'Agenda 21', aimed at promoting sustainable development throughout the world
- Tourism is recognised as an economic sector that needs to develop sustainably at 'Earth Summit II' in 1997 in New York.

The inherent ambiguity of the notion is reflected in organisations, the private sector, and academics, some of whom might be seen to have different and politically opposed purposes. This uncertainty enables different viewpoints on sustainability to be adopted. The Brundtland Report's most frequently cited definition of sustainable development is to blame for a large portion of this misunderstanding.

According to Richardson (1997), this definition is a political fudge designed to appease all parties involved by conceding the divergent opinions of commissioners from various states. The Brundtland's remaining portions, nevertheless.The report makes it clear that in order to fully implement this concept, it is necessary to address several critical development-related issues, such as reducing poverty, halting environmental deterioration, and concerns about equity across generations.

Population pressure has made poverty worse in many places, as the world's population grew quickly throughout the 20th century. The population of the world was 1,600 million in 1900, 6,000 million at the start of the new century, and the UN predicts that by 2025, there will be over 8,500 million people on the planet (World Guide, 1997).

Different Perspectives on Sustainable Development
Given the variety of objectives, interests, beliefs, and philosophies that underlie human engagement with the environment, it is probably not strange that there should be varying viewpoints on what sustainability means the surroundings. The terms "technocentrism" and "ecocentrism," which refer to two main ideologies towards the environment, are easily recognisable. Technocentrism is characterised by the reliance it has in quantifiable solutions to problems and the conviction that technical solutions to environmental problems may be found through the application of science. Because of its focus on quantification, it enables a detached objectivity in decision-making, disqualifying subjective elements of the environment like feelings or emotions. Such emphasis on objective measurement makes it possible to ignore the environment's complexity as a system and different points of view. According to Reid (1995: 131)

Because most development decisions have environmental effects that go beyond the immediate area in which they are made, there is also a lack of understanding of the complexity of the environment as a system.physical restrictions placed on a particular project. The physical world is seen as a resource that humans can use anyway they see fit according to the technocentric viewpoint. According to Pepper (1993: 34), technocentrism is characterised by a preference for centralised control over local decision-making. "There is little desire for meaningful public engagement in decision-making, especially to the right of this ideology, or for arguments about values," he adds.

Ecocentrism is a different ideological perspective on how we regard the environment (O'Riordan, 1981). Ecocentrism is characterised by a belief in the wonder of nature and is strongly related to the philosophical views of the romantic transcendentalists. Ecocentrics support alternative technologies as a solution because they lack faith in both current technology and the technological and bureaucratic elites. Alternative technologies are more democratic in that they can be owned, maintained, and understood by those with little economic or political power, in addition to being more likely to be ecologically friendly. The technological stance of ecocentrics is sometimes referred to as "Luddite," which means that while they are not opposed to new technology, they are opposed to technology that positions its ownership.

➢ That all beings, whether human or non-human, possess an intrinsic value, the antithesis of the technocentric instrumental viewpoint of nature
➢ That all beings are of equal value and there is therefore no hierarchy of species in nature
➢ That all nature is interconnected, with no dividing lines between the living and the non-living, the animate and inanimate, or the human and non-human
➢ That the earth is finite in its carrying capacity

Different Approaches to Development between the 'Dominant World-view' and 'Deep Ecology'	
Dominant world-view	**Deep ecology**
• Strong belief in technology is for progress and solutions	• Favours low-scale technology that self-relian
• Natural world is valued as a and resource rather than possessing intrinsic value	• Sense of wonder, reverence moral obligation vis-a`-vis the natural world
• Belief in ample resource reserves	• Recognises the 'rights' of nature are independent of humans
• Favours the objective and quantitive	• Recognises the subjective such as feelings and ethics
• Centralisation of power	• Favours local communities and localised decision making
• Encourages consumerism	• Encourages the use of appropriate technology
	• Recognises that the earth's resources are limited

Sustainable Consumption and Production

An strategy that is more closely associated with ecocentrism than with technocentrism occupies the third rung of the ladder. According to "strong sustainable development," environmental conservation should be a priority. a prerequisite for economic growth. As a result, in this viewpoint, environmental factors take precedence over economic growth, which was the case in the previous two scenarios. This viewpoint mandates that development strategies try to preserve other environmental assets deemed deserving of protection as they are, such as tropical rainforests, while maintaining the productive capability of environmental assets. Environmental quality is taken into account in "strong sustainable development," and "local communities" will participate in decision-making about development-related concerns. To support this strategy, all available policy instruments, including judicial, fiscal, and economic measures, should be utilised and modified.

What Baker refers to as the "Ideal Model" is at the top of the ladder. This strategy is supported by a strong ethical component, as it is believed that nature and non-human life have intrinsic value that goes beyond what humans can gain from them. As the "quality of life" rather than the "standard of living" is the goal of development, the quantitative measurement of growth becomes irrelevant. The policy ramifications of this point of view are that environmental conservation will severely curtail human activity, including economic activity, and resource use. This strategy for sustainable development entails systemic adjustments to the world's society and economy. Moreover, Baker claims that ecologists would contend that the "Ideal Model"

The argument for sweeping social reforms to promote sustainable development focuses on the balance of power in the larger political economy. This entails tackling the underlying factors that contribute to non-sustainability, such as the aforementioned issues of power and wealth distribution, the functions of multinational businesses, class-based politics, and gender inequality. Even though the Brundtland Report identified the distribution of intra-generational equity as a problem with direct implications for poverty, this topic is conspicuously absent from the majority of government agendas. According to Doyle and McEachern (1998: 37), "Radical

161

environmental political theorists are involved in paradigm struggles, each seeking to create new sets of key values and principles that directly challenge existing, powerful paradigms." Radical approaches to sustainable development go against the values and tenets of capitalist society.

According to Roussopoulos (1993), eco-feminism relates the world's current environmental issues to the patriarchal system and has its roots in antimilitarism. As a result, there is a connection between environmental damage and the treatment of women. With regard to eco-socialism, there is a large a variety of divergent political viewpoints, including socialist, liberal, social, and cultural ecofeminisms. So, a large portion of the discussion concerning how to define sustainable development is motivated by political tension. There is a fundamental divide between those who see sustainability as merely advancing technology and environmental accounting systems while maintaining the status quo of the social hierarchies and power structures, and those who have more radical political agendas that involve shifting the value systems.

Conclusions :

- Sustainable development is an approach to economic growth that places a strong emphasis on resource conservation and human development. The environment cannot be maintained if development doesn't account for the cost of destroying it, nor can it be developed upon a base of declining environmental resources. The primary goal of sustainable development is poverty reduction, or meeting the requirements of the global population without endangering the resources of the planet or the capacity of future generations to meet their own needs. As a result, it combines the concepts of equity between and among generations.

- The phrase "sustainable development" is imprecise and open to several interpretations by parties with divergent political and philosophical philosophies. Some ideologies and groups contend that fundamental social reform, including adjustments to political institutions and value systems, is the only way to achieve sustainable development. Others contend that technological advancements and changes to the market structure that do not endanger the status quo in society can lead to sustainable growth.

- Different viewpoints on the significance of the idea of sustainable development have also resulted from its application to the tourism industry. A majo.There is a distinction between those who view "sustainable tourism" as supporting the continuation of travel to a particular location and those who see travel as a means of promoting sustainable development. which covers a considerably broader range of socially decided objectives and aims. Three broad traditions of sustainable tourism are "resource-based," "activity-based," and "community-based," according to Saarinen (2006).

References
Baker, S., Kousis, M., Richardson, D. and Young, S. (eds) (1997) *The Politics of Sustainable Development: Theory, Policy and Practice within the European Union*, London: Routledge.
Doyle, T. and McEachern, D. (1998) *Environment and Politics*, London:Routledge.
Mowforth, M. and Munt, I. (1998) *Tourism and Sustainability: New Tourism inthe Third World*, London: Routledge.
Pepper, D. (1993) *Eco-socialism: From Deep Ecology to Social Justice*, London: Routledge.
Reid, D. (1995) *Sustainable Development: An Introductory Guide*, London: Earthscan.
Saarinen, J. (2006) 'Traditions of Sustainability in Tourism Studies', *Annals ofTourism Research*, 33 (4): 1121-40.

**Assistant professor & Head,
Dept of Geography,
Azad Mahavidyalaya AUSA Dist. Latur**

17. Conserving Natural Resources of Aggregates by using Demolished Coarse Aggregate

Dr. Chetan Khemraj[#]

Mrs. Sushma Barahate[*]

Abstract

Aggregates are the main material in concrete. The function of aggregate in concrete is to serve as filler. Aggregate give bulk to the concrete, strength, durability to concrete, decrease in shrinkage of concrete and achieve economical aspect. The main reality is that the aggregates have 70%-75% of the total amount by weight of a concrete mix. Various characteristics like strength of the concrete mix developed is liable to the property of aggregate used in the concrete mix. The strength of concrete mix design is also depends on the bond strength between the cement paste and aggregates. As the infrastructure industries increasing rapidly, the need of coarse aggregates are also rising. It is difficult to yield too much volume of coarse aggregate at this time. Therefore, there is a need rising for an substitute of coarse aggregates. Land filling is practiced by demolished concrete in the entire world but if recycling of demolished coarse aggregate is done then it may probably be an substitute to the coarse aggregate.

The main objective of this study is to sustain the natural resources of coarse aggregate by using demolished coarse aggregate. Experimental studies and analysis has been done on the strength characteristics and durability by replacing coarse aggregate by recycled coarse aggregates in high strength concrete. The tests are performed for strength and durability such as test for saturated water absorption, compressive strength test of cubes, acid resistance test and porosity. We replaced the coarse aggregates in M40 concrete mix by percentage of recycled coarse aggregate i.e. 0%, 10%, 20%, 30%, 40% and 50%. A sample of reduced water/cement ratio was also tested with 50% of recycled aggregate. From tests and analysis it was observed that RCA can be used in HCS by adjusting the water-cement ratio and contents of admixture in the concrete mix.

The experimental tests focused on physical properties of concrete, workability, density and mechanical properties and compressive strength of concrete with RCA. The compressive strength of RCA concrete with 60 % recycled aggregate, is about 76 % of natural aggregate concrete. As per study the amount of recycled coarse aggregate is inversely proportional to the compressive strength of concrete i.e. compressive strength decreases with increase in RCA.

Keywords : RCA, w/c ratio, workability, compressive strength, durability.

Introduction

At construction site we need several materials such as cement, concrete, aggregate, steel, brick, sand, stone, clay, glass, admixture, clay, wood, mud, and so on. However, the main construction material used at construction site is cement concrete. For its flexibility and suitability with respect to the changing environment conditions, the concrete mix must be such that it can conserve natural resources, save the environment, and create proper usage of energy. To accomplish this, crucial attention must be laid on the use of wastes products and by products in cement and concrete mix used for new constructions. As we know concrete comprises of 75% of aggregate, so use of demolished aggregate will play a major role in conserving natural aggregate. There is a lot of application of demolished aggregate in the construction industries.

Due to urbanisation, the construction work increase with rapid rate which leads to shortage of natural aggregate. So it's important to research on the usage of waste construction materials. With increase in construction and demolition practices, material waste is increasing which leads to pollution. The reasons behind the exploration and analysis is that demolished aggregate are simple to obtain and the cost of demolished aggregate is less than natural aggregate. With increase in construction need of aggregate is also increasing. Since there is limited source of natural aggregate, it is important to find out an alternative for natural aggregate. As demolished concrete is only used for land filling, it is better to demolished it for aggregates to reduced the need of natural aggregates.

Mechanical Properties
Compression Test Result and Analysis :
From the compression strength test it is noted that there is increase in the compressive strength of concrete cubes in the early period of time. However, it is also noted that the compressive strength of natural aggregate cube specimens is more than the demolished aggregate cube specimens. The graphical representation of deviation of compressive strength of concrete mix is showed in Figure 1.

The target strength to be achieved in this study is $40N/mm^2$. From the result obtained, it is noted that the only one batch fulfil the criteria of goal strength, which is the batch with 0% demolished aggregate. The compressive strength for remaining batches is less than $40N/mm^2$. However the compressive strength of the batch of 50% demolished aggregate with 0.35 w/c ratio is $37N/mm^2$, which is approximately equal to the target strength. So according to this, in case of demolished aggregate up to 30 to 40 % replacement may achieve target strength or high strength by reducing the w/c ratio.

Table No : 1 - 7th Day Compressive Strength Result in N/mm^2

Demolished Aggregate Used (%)	Compressive Strength in N/mm^2 after 7 days
0%	28.0
10%	27.1
20%	25.0
30%	24.2
40%	22.2
50%(0.40 w/c ratio)	19.0
50%(0.35 w/c ratio)	27.1

The initial compression strength of the mix having 50% of recycled aggregate with 0.35 w/c ratio is maximum, but after day 7, the rate of increment of compressive strength is decreasing when compared to other batches. It is also observed that with the increase in percentage of recycled aggregate, compressive strength of concrete specimen decreases. Figure 1 shows the initial compressive strength is maximum for specimen having 50% recycled aggregate with 0.35 w/c ratio.

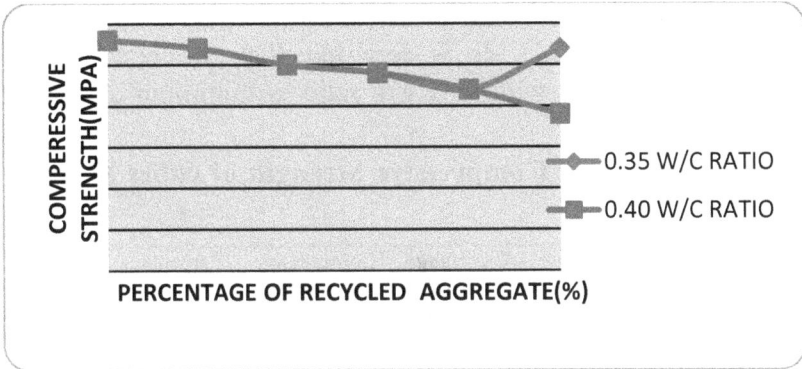

Figure 1: Deviation in compressive strength after 7 days of curing

Table No 2   - 28th Day Compressive Strength Result in N/mm^2

Demolished Aggregate Used (%)	Compressive Strength in N/mm^2 after 28 days
0%	41.0
10%	38.1
20%	35.0
30%	33.2
40%	30.0
50%(0.40 w/c ratio)	26.1
50%(0.35 w/c ratio)	38.0

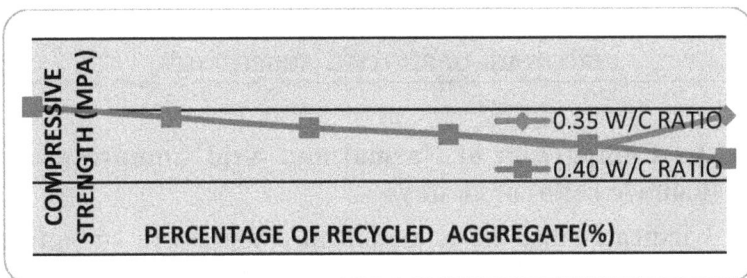

Figure 2 Deviation in compressive strength after 28 days of curing

Acid Resistance Test :

Table no.3 shows the results of acid resistance test (durability test). If the percentage of demolished coarse aggregate is increase, the loss of weight of concrete cubes also increase after 45 days of immersion in sulphuric acid (3%). It is reported that the concrete with demolished aggregate with less w/c ratio is minimum affected by sulphuric acid.

Table:-3 Decrease in Compressive Strength of cubes after Acid Resistance Test

Percentage Replaced of Aggregate	Compressive Strength in N/mm² after 28 days	Percentage decrease in weight of cubes	Compressive Strength in N/mm²	Percentage decrease in compressive strength of cubes compared to 28 day strength
0%	41.0	0.44	39.0	4.87
10%	38.1	0.47	34.7	89.23
20%	35.0	0.53	32.0	8.57
30%	33.2	0.56	28.3	14.75
40%	30.0	0.58	25.7	14.33
50%	26.1	0.65	20.4	21.83
50 % with decreased w/c ratio	38.0	0.54	32.3	15.00

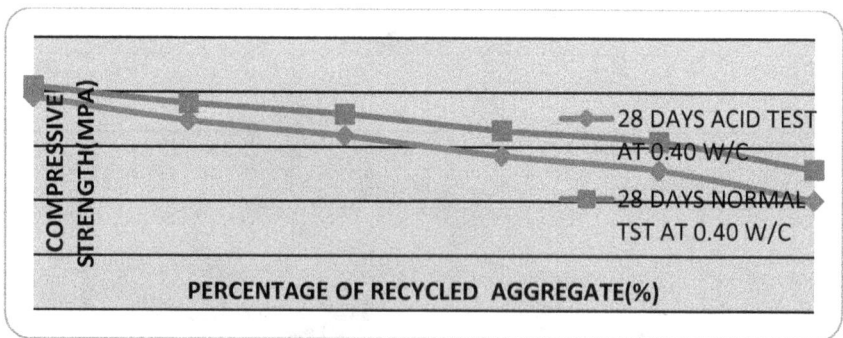

Figure 3 - Comparison of Normal and Acid Compressive Test Result (0.40w/c ratio) of 28 days

Figure 3 indicates that the reduction in compressive strength after 28th days is linear with the demolished aggregate. This is for the 0.40 w/c ratio for normal test and acid test for the specimen. The fig. 4 indicates that the 0% to 50% demolished aggregate compressive strength decreases for 0.40 water/cement ratio. But when the water/cement ratio reduces the strength of cube specimen is increase for 50% of demolished aggregate. The change of deviation show in fig. 3 & 4 for 50% demolished aggregate in different water/cement ratio. That is the reaction of acid in the 45 days in compressive strength result.

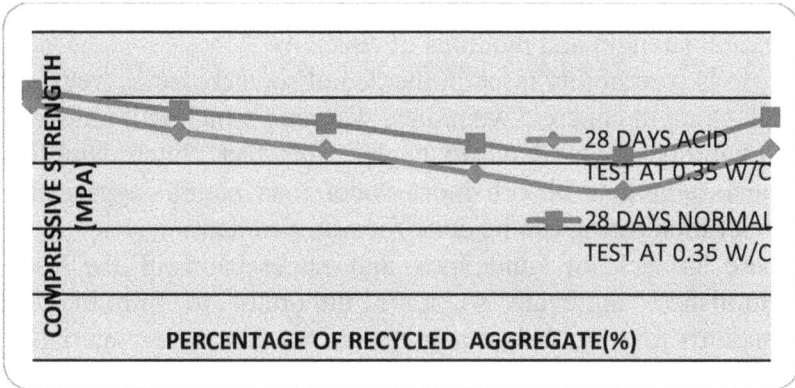

Figure 4- Comparison of Normal and Acid Compressive Test Result (0.35w/c ratio) of 28 days

Conclusion

Experimental studies and analysis had been done on the strength characteristics and durability by replacing coarse aggregates by demolished coarse aggregates in high strength concrete. The result of study are listed below :

1) With the increase in the % of demolished aggregates, compressive strength will decrease. But with reduction in water/cement, the compressive strength increases.

2) Decreasing the w/c ratio and mixing admixture in proper ratio will help to achieve the target compressive strength (40MPa) for 30 to 40 % of demolished coarse aggregate replacement. This is categorised as high strength performing concrete and can be used in infrastructures, which require compressive strength up to 40MPa.

3) After conducting the acid resistance test, the percentage loss in weight of concrete cubes is negligible for 30 to 40% recycled coarse aggregate replacements. It shows the cube mix are less attacked by acid.

4) In case of demolished coarse aggregate, the water absorption and porosity of cubes mix are higher than normal mix but within the permitted limits. By reducing water cement ratio and mixing admixtures, porosity and water absorption can be modified.

5) The research and analysis shows that demolished coarse aggregate can be used in high strength concrete mixes with suitable engineering properties such as compressive strength, flexural strength and modulus of elasticity.

6) There is workability issue in the demolished coarse aggregate, for high strength concrete we reduce w/c ratio which will reduces the workability, so we have to neglect this fact. Since demolished coarse aggregate absorb more water than natural aggregate, so proper monitoring can be done for water content.

7) There is lack of guidelines and supervision in the use of demolished aggregate so it is important to introduce new standards for demolished aggregates. So that these materials can be used successfully in practice, under arrange of environmental conditions.

References

[1] IS: 516-1959,Methods of test for strength of concrete, edition 1.2(1991-07).

[2] IS: 383-1970 (Second Revision), Specifications for Coarse and Fine Aggregates from Natural Resources for Concrete

[3] RILEM. "Specifications for concrete with recycled aggregates". Mater Structure, 1994, 27: 557–559.

[4] IS: 456 – 2000 (Fourth Revision) Indian Standard Plain and Reinforced Concrete Code of Practice.

[5] J.S.Ryu, (2002), "An experimental study on the effect of recycled aggregate on concrete properties", Mag. Concr. Res, Vol:54 (1), pp-7-12.

[6] Hong Kong Housing Department, Use of Recycled Aggregate, viewed 25 March 2004,

[7] Logic Sphere, Slump test, viewed 31 March 2004,

[8] CRISO, Australia First for Recycled Concrete, viewed 4 April 2004,

[9] M.S Shetty "Concrete Technology", Theory and practice, S Chand 2011.

[10] Butler L., 2012, "Evaluation of Recycled Concrete Aggregate Performance in Structural Concrete", thesis, The University of Waterloo, Canada, Viewed April 2013

[11] "An introduction to recycled aggregate concrete: Production and applications" (2015) by Piyush Sharma, Amity University.

Professor,

Sri Balaji College Of Engineering and Technology,

Jaipur

chetan.khemraj@gmail.com
*Research Scholar,
Mewar University, Rajasthan

18. ROS Generation Mechanisms as Antifungal Treatment : Potential Nanotube Delivery of Pyocyanin and Antioxidant Enzymes to Affected Tissue

Monisha Krishnan[1],
Parthiban R.[1],
Kiruthiga P.[1],
Nandhini R.[1],
Jagadeeswari M.[1],
Balachandar S.[1],
Devakumar J.[2]

Abstract

Reactive oxygen species (ROS) are formed upon reduction and oxidation of sensitive oxygen compounds, a cytolytic mechanism employed by microorganisms during pathogenesis. The compound pyocyanin is a phenazine pigment produced by *Pseudomonas aeruginosa*, which takes part in redox reactions and produces ROS, leading to antibacterial and antifungal activity. The present review considers the potential for pyocyanin to be used therapeutically through either internalization of pyocyanin in the fungal cells or slow release from within nanotubes that can use phages for attachment to the fungal cell wall. The targeted release of ROS will toxify the environment within and around the fungal cells, allowing for destruction of the fungal mycelia. The main challenge is damage to the host cells, which can be overcome by the use of superoxide dismutase, glutathione, and catalase enzymes at the infection site. The use of fungal phages will allow for targeting of the fungal cells for localization of the nanotubes. Such a treatment offers new hope in the fight against fungal infections since several are intrinsically resistant to known antifungal agents and often require surgical intervention in the course of treatment.

Keywords : antifungal resistance, nanotechnology, drug delivery, oxidative stress, antioxidants

Introduction

The threat of antifungal resistance is an increasingly large concern around the world. The risk increases when the immune systems of patients with comorbidities are compromised and they are able to acquire secondary infections, including fungal infections. Fungi most commonly encountered in the case of resistant infections include yeasts, such as *Candida* spp., and *Cryptococcus* spp., and molds, such as aspergilli, which have risen in mortality rates. Several fungi show intrinsic resistance to known antifungal agents while others have been observed to develop resistance over time due to selective pressure conferred by antifungal drugs (Fisher et al., 2022). The use of nanotechnology is being explored for greater effectiveness, specificity, and targeted impact.

Antifungal Resistance : An Emerging Concern

Fungi show varying degrees of resistance to antifungal agents, through either intrinsic or mutational mechanisms. Upon closer study of the yeast model, the commensal *Candida albicans* shows ability to invade mucosal tissues in cases of severe infection. A common clinical manifestation is oral thrush, the formation of a white layer on the surface of the tongue. It was observed that E-cadherin in the oral epithelial cells is proteolytically degraded by the yeast, facilitating the formation of deep-seated tissue infections in the body, particularly around the buccal cavity.

Aspergilli constitute a major particular class of molds causing infections with mycelial growth. They cause a number of manifestations in the human body, with invasive aspergillosis showing a mortality rate above 85%. *A. fumigatus, A. terreus,* and *A. niger* are commonly found, yet the resistance has been attributed to the difficulty in treatment and delayed diagnosis more than development of resistance over time, though they show varying degrees of intrinsic resistance to different azole agents, including triazole agents. The alternative options for treatment are predominantly intravenous drugs that are 15-20% less effective compared to azoles in treating invasive aspergillosis, where mortality was 20-30% higher than in patients with voriconazole resistance than sensitivity (Denning, 2022).

Mucor spp., (Mucorales) cause mucormycosis, a condition with a high mortality rate spanning from twenty to one hundred percent. Certain members, such as *Rhizopus oryzae*, show greater resistance to the innate host defense and thus have increased mortality rates compared to *Candida* and aspergilli. Mucormycosis is characterized by advanced angioinvasion that leads to thrombosis and tissue death; the former causes pathogenic distribution through circulation while the latter blocks immune responses to the site of infection. The predisposing factors include diabetic ketoacidosis and corticosteroid use; the fungus caused severe secondary infection in COVID-19 patients across India in 2019 and the main mode of infection was inhalation of the sporangiospores that led to tissue infarction and necrosis (Sannathimmappa et al., 2022). Rhino-orbital-cerebral mucormycosis is a manifestation requiring surgery that must not be sparing; both infected necrotic tissues and surrounding tissue must be removed due to the extent of Mucorales hyphal growth (Skiada et al., 2018); it is used in combination with amphotericin B. Posaconazole or isavuconazole is used for salvage therapy (Sannathimmappa et al., 2022).

ROS Generation

The body has a careful balance of oxidants and antioxidants, in which oxidants, such as free radicals, lead to oxidation of DNA, proteins, and heavy cell damage. Therefore, the use of ROS production as an antifungal agent should minimally expose host cells to avoid damage in the patient. Terbinafine, itraconazole, and amphotericin B are well known drugs which induce fungal generation of intracellular ROS upon exposure, a universal mechanism of drug action.

Surgery is an option that is used in conjunction with amphotericin B, yet it is of limited capacity in the cases of disseminated infection and when complex organs are infected, such as the brain. Therefore, the antifungal techniques have been mechanized to advanced levels for greater efficacy, where pressurized hyperoxygenation is a specialized technique. The method transfers significantly greater levels of oxygen through the blood and into the tissues, allowing for inhibition of anaerobic growth. The use of hyperbaric oxygen has been studied *in vitro* and in vivo, where *A. fumigatus* has been a model organism, showing reduction of biofilm formation in vitro and a dose-dependent effect in vivo (Dhingra et al., 2018). The function of fungal superoxide dismutase genes was lost, extending the effects of the therapy. Hence the intensity of oxygen concentration is a significant factor in destroying fungal cells.

The significance of high oxygen intensities is observed, yet the effect of oxidative stress at the cellular level is an important observation associated with dysfunctional cellular metabolism. ROS production by fungal cells is found to be enhanced upon exposure to stresses, including starvation, light, antifungal agents, and temperature (Gessler et al., 2007). Moreover, intracellular ROS, such as hydrogen peroxide and superoxide, are produced by the mitochondria as byproducts of cellular metabolism. Antifungal agents tested for inducing intracellular ROS generation in *A. fumigatus* were thymol, farnesol, citral, nerol, salicylic acid, phenazine-1-carbonic acid, and pyocyanin (Oiki et al., 2022). *A. oryzae* has shown reductions in biomass upon increased concentrations of oxidants, such as hydrogen peroxide (Shao et al., 2019)

The use of hyperbaric oxygen is an effective technique, yet the precision involved in nanolevel targeting enhances efficiency of the mechanism and minimal effect on the host system when executed properly. This precision is the principle and purpose behind nanotechnological approaches, including dissemination of nanoparticles in a specified site and drug delivery mechanisms involving affinity binders or biomarker recognition.

Pyocyanin Production and Potential as Antimicrobial Agent

Pseudomonas aeruginosa is an aerobic, Gram-negative bacterium that is known for rapid development of resistance to standard antibiotics, allowing it to run rampant in healthcare settings as a nosocomial pathogen. *P. aeruginosa* is found commonly in soil, water, and on the surfaces of a number of objects. It thus fulfills an important niche in the environment and pyocyanin is a regularly observed metabolite first found in the form of blue-colored pus oozing from the infected open wound of a patient. The pathogen is opportunistic and thus invades in the case of immunocompromised and immunosuppressed patients. It produces aeruginosin A and B, pyocyanin, pyoverdine, pyorubin, and other variant pigments (Muller and Merrett, 2014). Pyocyanin (5-Methyl-1(5H)-phenazinone) is a blue-green phenazine pigment that is an electron shuttle known to enhance the survival of the bacterium in a biofilm. It turns red upon reduction in pH and blue at higher pH values. Unlike pyoverdine, pyocyanin has not been directly associated with siderophore roles, thus playing minimal role in the fungal mechanism of iron sequestration which supplements its survival. The incorporation of pyoverdine may enhance the siderophore capability by attracting the iron particles to the nanotube rather than fungal site tissue where several fungi can assimilate the iron molecules (Sass et al., 2021).

Pyocyanin, while associated with damage to host tissue, is a well-studied metabolite that has been tested in encapsulated nanoparticle form against several fungi, leading to reductions in growth. In relation to fungal infection of seedlings, it was observed that pyocyanin inhibited fungal growth at higher concentrations whereas lower concentrations increased colony-forming units on seedlings (Khare and Arora, 2011). Moreover, non-limiting conditions provide an important observation: when iron is not a limiting factor, pyoverdine is not found, yet pyocyanin is produced, as in the case of blood infections by *P. aeruginosa,* where hemoglobin provides great iron content (Sass et al., 2021). Pyocyanin clearly shows inhibitory activity against *Candida* spp., and *A. niger* (Mrrez and Haitham, 2020). The role of pyoverdine in iron sequestration should be estimated in significance, in order to block its siderophore capabilities that enrich the growth of aspergilli or shift its assimilation from the site of primary infection.

Pyocyanin can be applied to effluents that leave hospitals and pharmaceutical industries, as it acts as both a redox and pH indicator, thus indicating successful biotransformation of any bio-augmented bacteria. More importantly, the role in treating bacterial infections may be minimal compared to antifungal effect. It is already compared to fluconazole for its efficacy in treating *T.*

rubrum superficial infections. The production of ROS is harmful in the case of eukaryotic and prokaryotic cells, yet catalase minimizes radical generation in a number of cases. In regard to antibacterial activity, pyocyanin is effective against bacterial cells with catalase as well. It often exhibits antimicrobial activity adequate to allow for isolated growth, enhancing the biofilm-forming ability of *P. aeruginosa* as well. Its antifungal activity has also been tested at different concentrations to great effect.

Nanotube Delivery Approach

Nanodelivery involves the use of materials with sizes between 1 to 100 nm and interactions at the nanolevel for greater precision in targeting and reduction of host exposure. Mechanisms include nanospheres or nanotubes for delivery through encapsulation, nanoparticle coatings, or filling of nanotubes. Nanostructures have also shown prolonged presence in blood circulation and allow for periodic releases of the agent over time (Patra et al., 2018). Nanotubes offer dual functionality in that they are able to be coated with targeting molecules or particles that help evade the host response (Goldberg et al., 2007). Pyocyanin nanoparticles have already been prepared and utilized for antifungal coatings which showed stability for nearly three months, as well as logarithmic reduction of *Aspergillus niger* (da Silva et al., 2022). For treating skin disorders, the prospects of nanodelivery for topical agents has been explored, focusing on liposomes and polymeric nanoparticles as molecules to penetrate the stratum corneum into the epidermis and dermis layers for relief (Cui et al., 2021).

Carbon nanotubes (CNTs) are well known compared to other material compositions. Their features offer functional advantages, including high surface area host system-drug contact and capacity for drug loading, stability, a syringe-like shape for inserting into cell walls, and compatibility with host microenvironments (Zare et al., 2021). They also allow for targeted release of the agent. CNTs do not have a biodegradability mechanism which is yet to be developed. Their applications span bioimaging, gene therapy, and other localized challenges (Cooper et al., 2015). CNTs have proven to promote the antifungal efficacy of Nystatin by acting as functionalized nanovectors, as they are able to easily cross cell membranes into the fungal cells (Uttekar et al., 2016).

Hydrogel nanotubes have been developed for diabetic wound healing. Microgranules are normally hydrogel-based vesicle-like structures that carry a compound and are used for local delivery of a compound. They have been used when incorporating clotrimazole for delivery to fungal infections with a sustained release being observed upon use. The delivery of microgranules containing pyocyanin is a new type of solution in that the ROS will overwhelm the cell metabolism and structure, leading to lysis. Unlike other agent-based treatments, there is not a possibility for immediate development of resistance to a chemical compound.

CNTs present the possibility for targeted delivery through attachment of functional components, including antigens, antibodies, and bacteriophages. Antigens were used for the ability to target melanoma cells, resulting in inhibition of metastasis through antigen mimetics. Furthermore, the binding of phages to nanotubes as conjugates can be done cost-efficiently and while maintaining viability of the phages, as done for magnetically coupled bacteriophages for direct antimicrobial activity toward pathogens, particularly using phages against *P. aeruginosa*, which incidentally show antifungal activity as well. Such conjugates, while also effective toward the pathogens themselves, can be used for specificity of attack using affinity and the CNTs can be allowed to release chemicals at the site once bound to the pathogen through phage attachment, either serially or in a concentrated manner at a site of advanced infection, such as that indicated by heavy mycelial growth or patches of spore germination detected by skin or tissue biopsy.

The cytoplasmic streaming within hyphae offers an advantage in mold treatment, as a drug or agent can be dispersed throughout a network of hyphal filaments by crossing through septal pores in the walls separating hyphal compartments (Bleichrodt et al., 2015).

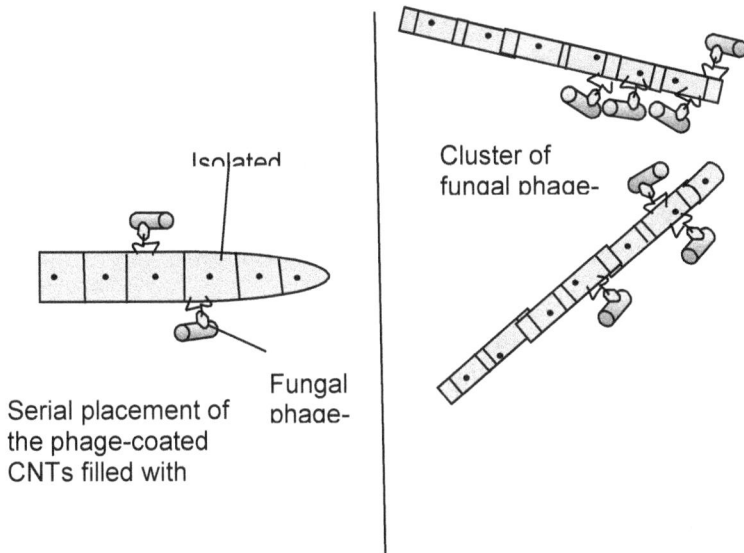

Figure: Potential mechanisms for specificity to the fungal mycelium.

Different variations can be made at various steps of the process :

i. Attachment

Phage-coated CNTs can be made to surround the mycelia by bacterial antigen affinity or chemical methods for activation mechanism if needed for delivery. However, the use of replication-controlled fungal phages is adequate for the phage-CNT conjugates to attach to their specific targets.

ii. Release of Pyocyanin

Location : CNTs can release the pyocyanin at the site of infection, either into the mycelia or around the hyphae, into the surrounding microenvironment. The needle-like structure of the CNTs allows for shape-selective attachment, insertion, and release of the pyocyanin within the hyphae.

Dissemination : When the CNTs are internalized, the pyocyanin released can cause rapid accumulation of ROS and cause cytolysis from the inside out. The number of CNTs required for hyphae can be altered based on the effectiveness of the dissemination of pyocyanin; further study is required to determine whether septal pores can serve as internal channels for spreading of pyocyanin along the length of the hyphae. If such a mechanism occurs, attachment at a single site on the hyphae may be sufficient. Intercompartmental cytoplasmic streaming in hyphae must be utilized for dispersing of the agent throughout the mycelia in advanced patients.

Frequency : Either slow or periodic release of pyocyanin can be explored, as well as a singular instance of release.

Enzymatic Protection

While pyocyanin provides a physical mode of antifungal treatment, host cells in the surrounding infected tissue may also be damaged by the ROS produced. For this, either preliminary or subsequent enzyme therapy may be effective in protecting the host cells, since the fungal cells are separate in the microenvironment. Enzymes include those that are antioxidant defenses, such as glutathione peroxidase, superoxide dismutase, catalase, and peroxiredoxins. These enzymes prevent tissue damage caused by ROS by neutralizing free radicals, including hydroperoxides and hydroxyls (Ighodaro and Akinloye, 2017).

The highest tier of antioxidants is comprised of glutathione peroxidase, catalase, superoxide dismutase, as well as transferrin which commonly convert hydrogen peroxide to more tolerable products such as water and oxygen. Metal ion binders also help prevent free radical formation by assimilating iron or other reactive metal ions, a mechanism that could help since competitive binding prevents fungal assimilation of the metal ions. The secondary tier consists of hydrophilic and lipophilic agents, such as ascorbic acid and ubiquinol, respectively. The third tier contains antioxidant enzymes for treating the damage, including DNA repair enzymes, such as polymerases, glycosylases, and nucleases, as well as proteases (Ighodaro et al., 2017).

Whether provided as a prophylactic or therapeutic agent, the enzymes should be delivered in a way that does not negate the ROS-based cell destruction mechanism itself.

Conclusion

Antifungal agents are innumerable, as plant extracts, microbial metabolites, and physicomechanical mechanisms are effective as treatment of severe fungal infections. However, the intrinsic resistance of significant pathogenic molds, such as *Aspergillus* spp, and *Mucor* spp., require surgical intervention and the use of potent agents that cause equal damage to both the pathogen and the affected patient, bringing forward the demand for newer agents that are more effective. Pyocyanin provides a novel option of utilizing ROS production as an antifungal mechanism, since the effectiveness of chemical agents varies for different fungal pathogens. The use of CNTs with pyocyanin can be enhanced through the incorporation of targeting molecules that provide a specific mechanism for affinity and attachment to the fungal cells. Further study of the chemical permeability through septal pores within fungal hyphae will augment the effectiveness of the therapeutic response. The introduction of pyocyanin as an antifungal agent in cases of severe infection will save the lives of numerous immunocompromised patients.

References
1. Bleichrodt, R. J., Vinck, A., Read, N. D. & Wösten, H. A. B. Selective transport between heterogeneous hyphal compartments via the plasma membrane lining septal walls of *Aspergillus niger*. *Fungal Genet. Biol.* 82, 193–200 (2015).
2. Cooper, I. R., Illsley, M., Korobeinyk, A. V., & Whitby, R. L. (2015). Bacteriophage-nanocomposites: an easy and reproducible method for the construction, handling, storage and transport of conjugates for deployment of bacteriophages active against Pseudomonas aeruginosa. *Journal of microbiological methods*, *111*, 111–118. https://doi.org/10.1016/j.mimet.2015.02.005
3. Cui, Mingyue & Wiraja, Christian & Chew, Sharon & Xu, Chenjie. (2020). Nanodelivery Systems for Topical Management of Skin Disorders. Molecular Pharmaceutics. XXXX. 10.1021/acs.molpharmaceut.0c00154.
4. da Silva, J. E. G., do Amaral, I. P. G., Kretzschmar, E. A. de M., & Vasconcelos, U. (2022). Antifungal Coating Based on Pyocyanin Nanoparticles (Np-Pyo). *European Journal of Biology*

and Biotechnology, *3*(2), 30–37. https://doi.org/10.24018/ejbio.2022.3.2.360

5. Dhingra, S., Buckey, J. C., & Cramer, R. A. (2018). Hyperbaric Oxygen Reduces Aspergillus fumigatus Proliferation *In Vitro* and Influences *In Vivo* DiseasA Skiada, C Lass-Floerl, N Klimko, A Ibrahim, E Roilides, G Petrikkos, Challenges in the diagnosis and treatment of mucormycosis, Medical Mycology, Volume 56, Issue suppl_1, April 2018, Pages S93–S101,e Outcomes. *Antimicrobial agents and chemotherapy*, *62*(3), e01953-17. https://doi.org/10.1128/AAC.01953-17

6. Gessler, N. N., Aver'yanov, A. A., & Belozerskaya, T. A. (2007). Reactive oxygen species in regulation of fungal development. *Biochemistry. Biokhimiia*, *72*(10), 1091–1109. https://doi.org/10.1134/s0006297907100070

7. Fisher, M.C., Alastruey-Izquierdo, A., Berman, J. *et al.* Tackling the emerging threat of antifungal resistance to human health. *Nat Rev Microbiol* 20, 557–571 (2022). https://doi.org/10.1038/s41579-022-00720-1

8. Goldberg, M., Langer, R., & Jia, X. (2007). Nanostructured materials for applications in drug delivery and tissue engineering. *Journal of biomaterials science. Polymer edition*, *18*(3), 241–268. https://doi.org/10.1163/156856207779996931

9. KhareEkta and AroraNaveen Kumar. Dual activity of pyocyanin from *Pseudomonas aeruginosa*—antibiotic against phytopathogen and signal molecule for biofilm development by rhizobia. *Canadian Journal of Microbiology*. 57(9): 708-713. https://doi.org/10.1139/w11-055

10. Denning D. W. (2022). Antifungal drug resistance: an update. *European journal of hospital pharmacy : science and practice*, *29*(2), 109–112. https://doi.org/10.1136/ejhpharm-2020-002604

11. Diaa A Marrez and Haitham SM. Biological activity and applications of pyocyanin produced by *Pseudomonas aeruginosa*. 2020 - 2(1) OAJBS.ID.000133.

12. Muller, M., & Merrett, N. D. (2014). Pyocyanin production by Pseudomonas aeruginosa confers resistance to ionic silver. *Antimicrobial agents and chemotherapy*, *58*(9), 5492–5499. https://doi.org/10.1128/AAC.03069-14

13. Oiki, S., Nasuno, R., Urayama, Si. *et al.* Intracellular production of reactive oxygen species and a DAF-FM-related compound in *Aspergillus fumigatus* in response to antifungal agent exposure. *Sci Rep* 12, 13516 (2022). https://doi.org/10.1038/s41598-022-17462-y

14. Sass, G., Nazik, H., Chatterjee, P., & Stevens, D. A. (2021). Under nonlimiting iron conditions pyocyanin is a major antifungal molecule, and differences between prototypic Pseudomonas aeruginosa strains. *Medical mycology*, *59*(5), 453–464. https://doi.org/10.1093/mmy/myaa066

15. A Skiada, C Lass-Floerl, N Klimko, A Ibrahim, E Roilides, G Petrikkos, Challenges in the diagnosis and treatment of mucormycosis, *Medical Mycology*, Volume 56, Issue suppl_1, April 2018, Pages S93–S101, https://doi.org/10. 1093 /mmy/myx101

16. Tragiannidis, A., & Groll, A. H. (2009). Hyperbaric oxygen therapy and other adjunctive treatments for zygomycosis. *Clinical microbiology and infection : the official publication of the European Society of Clinical Microbiology and Infectious Diseases*, *15 Suppl 5*, 82–86. https://doi.org/10.1111/j.1469-0691.2009.02986.x

17. Ighodaro, Osasenaga & Akinloye, Oluseyi. (2017). First line defence antioxidants-superoxide dismutase (SOD), catalase (CAT) and glutathione peroxidase (GPX): Their fundamental role in the entire antioxidant defence grid. Alexandria Journal of Medicine. 54. 10.1016/j.ajme.2017.09.001.

18. Patra, J.K., Das, G., Fraceto, L.F. *et al.* Nano based drug delivery systems: recent developments and future prospects. *J Nanobiotechnol* 16, 71 (2018). https://doi.org/10.1186/s12951-018-0392-8

19. Sannathimmappa, M. B., Nambiar, V., & Aravindakshan, R. (2022). Storm of a rare opportunistic life threatening mucormycosis among post COVID-19 patients: A tale of two pathogens. *International journal of critical illness and injury science*, *12*(1), 38–46. https://doi.org/10.4103/ijciis.ijciis_48_21

20. Shao, H., Tu, Y., Wang, Y., Jiang, C., Ma, L., Hu, Z., Wang, J., Zeng, B., & He, B. (2019). Oxidative Stress Response of

Aspergillus oryzae Induced by Hydrogen Peroxide and Menadione Sodium Bisulfite. *Microorganisms*, 7(8), 225. https://doi.org/10.3390/microorganisms7080225

21. Singh, Aarti & Bhattacharya, Rohan & Shakeel, Adeeba & Kumar, Arun & Jeevanandham, Sampathkumar & Kumar, Ashish & Chattopadhyay, Sourav & Bohidar, Himadri & Ghosh, Sourabh & Chakrabarti, Sandip & Rajput, Satyendra & Mukherjee, Monalisa. (2019). Correction: Hydrogel nanotubes with ice helices as exotic nanostructures for diabetic wound healing. Materials Horizons. 6. 10.1039/C9MH90006C.

22. Uttekar, Pravin & Kulkarni, Akshata & Sable, Pravinkumar & Chaudhari, Praveen. (2016). Surface Modification of Carbon Nanotubes with Nystatin for Drug Delivery Applications. Indian Journal of Pharmaceutical Education and Research. 50. 385-390. 10.5530/ijper.50.3.10.

23. Zare H, Ahmadi S, Ghasemi A, Ghanbari M, Rabiee N, Bagherzadeh M, Karimi M, Webster TJ, Hamblin MR, Mostafavi E. Carbon Nanotubes: Smart Drug/Gene Delivery Carriers. *Int J Nanomedicine*. 2021;16:1681-1706 https://doi.org/10.2147/IJN.S299448

[1]**Department of Microbiology,**
Rathnavel Subramaniam College of Arts and Science,
Bharathiar University, Coimbatore, India
[2]**Department of Microbiology,**
Dr. NGP Arts and Science College,
Bharathiar University, Coimbatore, India
Corresponding Author : Monisha K.,
email : monishak4477@gmail.com

19. Issues and Challenges in Conservation of Biodiversity in Protected Areas : A Case Study in Laokhowa Wildlife Sanctuary

Dr. Sanjeeb Kumar Nath

Abstract

Conservation issues and challenges in terms of biodiversity conservation in protected areas like National Parks, Sanctuaries, Conservation Reserves and Community Reserves a stronghold for biodiversity harbouring a variety of animal and plant species of economic, ecological and socio-cultural importance is a debatable topic. Major efforts to protect these resources have been undertaken against destruction and loss of resources of protected areas for proper management. However, these areas and its adjacent areas have long been subjected to a number of burning issues and challenges, which complicate proper conservation measures, thus putting the resources at risk of over exploitation and extinction. These issues and challenges include, among other things, government policies, failure of conservation to compete effectively with alternative land uses, habitat degradation and blockage of wildlife corridors, overexploitation and illegal resource extraction, wildfires, human population growth, poverty, and human-wildlife conflicts. In this paper, we review these conservation issues and challenges in biodiversity conservation by drawing experience from a case study of Laokhowa Wildlife Sanctuary of Nagaon district of Assam, India. This paper presented a framework on issues and challenges for discussions about ways to improve conservation and management that achieve the objectives, and at the same time promote local and national development, and contribute to sustainable local livelihoods. We conclude by recommending some stringent measures that may enhance the sustainability of the protected areas resources for the benefit of the mankind.

Keywords : Habitat degradation, invasive species, illegal hunting, wildfires

Introduction

Biodiversity loss due to habitat loss and deforestation, climate change, excessive nutrient load, unsustainable use, invasive alien species and human expansion activity (Harrison, S., Bruna, E, 1999, Debinski, D.M. Holt, R.D., 2000, Zipkin, E.F. DeWan, A. Royle, A.J. , 2009 and Moreno-Sanchez, R. Moreno-Sanchez, F. Torres-Rojo, J.M. 2011) researchers and decision makers across the world to think on the natural resource management and explore alternative approaches that are effective in preventing ecosystem degradation and species extinctions (Moreno-Sanchenz, R.,et.al 2012), and at the same time promote sustainable resource use. In recent years, many research studies have pointed out that the biodiversity loss has increased dramatically due to increasing human intervention in the natural environment (Vitousek, P.M, et.al 1997), and species are estimated to be disappearing at a rate more than a thousand times faster than is known historically (Dash, M. Behera, B.,2012). This loss of species threatens the availability of essential ecosystem services that are vital for the survival of human communities. In an attempt to tackle this situation, forest agencies across the globe have adopted conservation and management policies by creating wild habitats such as Protected Areas (PAs) in the form of biosphere reserves, national parks and wildlife sanctuaries, for biodiversity conservation that is critically endangered, threatened or vulnerable (Westing, A.H.,1998). The importance and relevance of PAs lies in the conservation of bio-resources, and also in supporting sustainable development initiatives. However, presently the PAs in India are facing numerous challenges and are in critical and threatened condition (Kothari, A. Suri, S. Singh, N., 1995). The local indigenous people lack their customary rights on land and park resources, which have raised fundamental issues about the survival of local communities in BR, national park and wildlife sanctuaries (Neuman, R.P. Machlis, G.E., 1989). This has an impact on the survival and livelihood base and refrain of local people for their basic inputs like non-timber forest products (NTFPs), firewood and fodder for livestock; causes eviction of local traditional communities by displacing people, and cutting them off from their principal source of economic livelihood and results in various environmental problems and socio-economic conflicts (Dowie, M., 2005). Local support through socio-economically beneficial activities such as tourism, alternative employment opportunities, cultural preservation and making local people essential shareholders in conservation benefits helps in PAs management. The present research focused on forest management within PAs considering the Laokhowa Wildlife Sanctuary as the main study area. Efforts are made to analyze the current biodiversity richness, threat status, current issues and future challenges in terms of conservation faced by the PAs of India. The study is mainly based on primary data collected during field study, secondary information collected from forest guards/officers, review of published literatures, and interactions with local people living in and around the sanctuary dealing forest management. The objective of the paper was to describe the richness of biodiversity, address the forest issues and challenges, current status, future challenges and ways to improve forest management.

Material and Methods

Stydy Area :

Laokhowa Wildlife Sanctuary is situated in the Nagaon district of Assam, India between the latitudes 26030 /N to 26032 / N and longitude 92040 / E to 92047 / E in the flood plains of the river Brahmaputra. The Sanctuary is about 25 km from Nagaon town, the district headquarter of the Nagaon district of Assam. It is located in the central part of the state of Assam and is situated in the extreme northern boundary of Nagaon district and the southern boundary of Sonitpur district. It is bounded by Burachapori Wildlife Sanctuary, Laokhowa suti, Haldia suti, and Mara suti in the north, Nagaon – Silghat PWD road in the east, Leterijan (water body) in the south and forest road in the west. Geomorphologically, the sanctuary consists of basically a flat land and the monotony of the plain is to a certain extent broken by the presence of wetlands (nallas, beels). The land has gentle slope from south to north and east to west .It is a part of Brahmaputra valley. The soil of the area is mostly alluvial deposits of the river Brahmaputra. Soil is generally fertile, clay loam mixed with silt.

Climate :

The climate of the sanctuary is characteristically monsoonal with a rhythm of changing season. It changes with respect to the changing climatic elements, which effectively controls the biodiversity of the area. The climate of the Laokhowa Wildlife Sanctuary can be treated as sub-tropical monsoonal type climate. Annual temperature of the sanctuary varies between 9.60C (min) and 33.80C (max). Average annual rainfall remains around 2000 mm and about 70% occurs during June – September. The relative humidity varies between 65 – 95% and is lowest during the month of March.

Methodology

The methodologies for the study were :

a) Intensive field visits in the villages. Field study with different user groups and knowledgeable individuals helped to understand different aspects related to biodiversity.

b) Information collection through interviews with the forest officials, forest villagers and the villagers living around the Sanctuary.

Conservation Issues and Challenges af Biodiversity Conservation in Laokhowa Wildlife Sanctuary

i) **Habitat destruction :** The natural habitat of the sanctuary has been destroyed mainly for timber and cultivation purposes. The major portion of the sanctuary has been destroyed mainly near Singimari and Laokhowa area of the sanctuary with intensions for permanent settlement and cultivation by the forest villagers.

ii) **Biotic Pressure :** Biotic pressure in the form of fishing, collection of wood for fuel and house building materials, collection of thatch, livestock grazing is very high in the sanctuary. Illegal fishing is very common in almost all the water bodies of the sanctuary. Since nearby villagers are poor, many of them are totally dependent on the sanctuary for fuel wood and other requirements. Hostile attributes, population explosion, extreme poverty of surrounding villagers are the major causes of biotic pressure on the sanctuary.

iii) **Erosion :** Annual periodic flood of the Brahmaputra river is another major problem of the sanctuary which causes erosion. In

188

2002, a major portion of the Roumari beel of the sanctuary silted up as embankment on the southern side was broken by the water of the Brahmaputra river.

iv) **Management Problem :** Many obnoxious weed species like Mikania micrantha, Chromolaena odorata etc. are gradually spreading all over the sanctuary there by causing retarded growth of many plant species which in turn is affecting the wildlife of the sanctuary.

v) **Staff :** As reported, the forest staffs of the sanctuary is not adequate and are also not well equipped to keep control over the sanctuary to make it free from biotic interferences. Many forest staffs are working on adhoc basis for many years with a very little remuneration. Many of them are physically assaulted by the villagers of neighbouring area, which has demoralised them.

Recommendations on Potential Solutions for Biodiversity Conservation in Laokhowa Wildlife Sanctuary :

1. Inside the Laokhowa Wildlife Sanctuary human activities are so intense that a few forest staff members cannot give enough protection to the sanctuary. The biodiversity of the sanctuary has great significance with its rich germ plasm. The concerned authorities should take strict conservation measures to protect the habitat and biodiversity of the sanctuary.

2. The forest villages and Taungya villages of the sanctuary should be acquired from the villagers and the displaced persons should be allotted suitable land elsewhere.

3. The government should appoint well equipped forest staff for better conservation works

4. All illegal activities such as encroachments, felling, uprooting of trees, grazing, grass and thatch collection and fuel-wood collection should be stopped.

5. Invasive weed species like Mikania micrantha, Eichhornia crassipes and Eupatorium odoratum etc. should be removed as far as possible for better protection of the grasslands and other vegetation areas.

6. All forest camps which are in dilapidated condition should be repaired and a few camps, mainly in between Lathiamari camp

and Laokhowa Beat which are strategically vary important for protection of the sanctuary.

7. People of nearby villages, Taungya villages and Forest villages and encroachers use pesticides in the agricultural fields, which are harmful not only to human beings but also for the other animals and plant species. Use of such pesticides in side and in the adjacent areas of the sanctuary should be restricted.

8. Regular census of the wild animals should be carried out.

9. It is necessary to educate the local people of the area about eco-friendly activities and also about the importance of conservation and to involve them in the various management activities of the sanctuary.

10. There is an urgent need to introduce eco-tourism in the area and socio-economic improvement programme to decrease human pressure on the resources of the sanctuary.

11. To introduce JFM (Joint Forest Management) programme in the sanctuary to reduce the pressure on the resources of the sanctuary.

Conclusion :

The protected area network in India has helped to conserve country's biodiversity. The network of PAs currently covers an area of 8.1 million ha, encompassing about 14 percent of the country's forest area and 4.61 percent of its land mass. From six national parks and 59 wildlife sanctuaries in 1970, the numbers increased to 85 and 462 in 1998, respectively (Wildlife Institute of India, 1998). According to a survey carried out in the mid-1980s, over 65 percent of the PAs were characterized by human settlements and resource use (Kothari et al., 1989).

The basic approach to management of PAs has been isolationist, based on the questionable assumption that certain areas are pristine or primary and that management must protect the park from people living in surrounding areas and shield wildlife and other natural resources from exploitation. The need and importance of exclusion of people from protected areas is in itself debatable. Numerous ecological studies have shown that not all human use is detrimental to wildlife conservation. Throughout the world, present-day forest quality and biodiversity patterns reflect the influence of past land use practices (e.g. Gomez-Pompa and Kaus, 1992). In fact, in some particular cases, it has been observed that excluding human activities from ecosystems can actually reduce biodiversity and lead to habitat deterioration (e.g. Hussain, 1996), while certain habitats have improved following human use/habitation (e.g. Ramakrishnan, 1992).

It has been found from the studies of the vegetation, that frequent floods and biotic disturbances like felling of trees, excessive grazing etc. led to a series of ecological problems, such as extinction of plant species, soil erosion, depletion of wildlife and other germplasm resources of the sanctuary. The management policies of the sanctuary should be aimed to preserve and protect the native flora and fauna of the sanctuary as they are directly interdependent on each other and help in maintaining ecological balances. Therefore, it is essential that the area be preserved involving everyone in general and local people in particular. The present management policies in the sanctuary should be improved to stop over exploitation of the resources viz., unauthorized collection of timber, over fishing, collection of wood for fuel and house building material, collection of thatch, livestock grazing.

Therefore it has become essential to conserve biodiversity in sizable natural areas for scientific, educational, ecological, recreational and economic development. The forest cover is decreasing day by day due to the human activities leading to ecological problems. Although there is no encroachment inside the sanctuary, but unfriendly villagers of surrounding areas who are poor and illiterate constantly destroying the habitat by way of forcible fishing, felling of trees. The various problems of the sanctuary and protection of the natural habitat of the Wildlife conservation of the flora and fauna should be given prime importance. This study would be helpful to the students, Researchers, Environmentalists, NGO and also the Foresters for proper management and conservation of bioresearches of the sanctuary.

Reference :

Dash, M. Behera, B. Management of Similipal Biosphere Reserve Forest-Issues and Challenges, Advances in Forestry Letters, 1(1), 2012, pp. 7-15.

Debinski, D.M.   Holt, R.D. A survey and overview of habitat fragmentation experiments, Conservation   Biology, 14, 2000, pp. 342-355.

Dowie, M. Conservation refugees: when protecting nature means kicking people out, Orion Magazine, 11/12, 2005, pp. 16-27.

Gomez-Pompa, A. & Kaus, A. 1992. Taming the wilderness myth. BioScience, 42(2): 271-279.

Harrison, S., Bruna, E. Habitat fragmentation and large-scale conservation: what do we know for sure? Ecography, 22, 1999, pp. 225-232.

Hussain, S.A. 1996. A case study on effective wetland management. Dehradun, India, Wildlife Institute of India. (unpublished mimeo)

Kothari, A. Suri, S. Singh, N.Conservation in India: A New Direction, Economic and Political Weekly, 30(43), 1995, pp. 2755-2766.

Kothari, A., Pandey, P., Singh, S. & Variava, D. 1989. Management of national parks and sanctuaries in India. Status report. New Delhi, India, Indian Institute of Public Administration.

Moreno-Sanchenz, R.   Torres-Rojo, J.M.   Moreno-Sanchez, F. Hawkins, S. Little, J. McPartland, S., National assessment of the fragmentation, accessibility and anthropogenic pressure on the forests in Mexico, Journal of Forest Research, 23(4), 2012, pp. 529-541.

Moreno-Sanchez, R.   Moreno-Sanchez, F.   Torres-Rojo, J.M. National assessment of the evolution of forest fragmentation in Mexico, Journal of Forest Research, 22(2), 2011, pp. 167-174.

Neuman, R.P. Machlis, G.E. Land use and threats to parks in the neotropics, Environment Conservation, 16(3), 1989, pp. 13-18.

Ramakrishnan, P.S. 1992. Shifting agriculture and sustainable development. Man and the Biosphere Series No. 10. Paris, UNESCO/Parthenon Publishing Group.

Vitousek, P.M. Mooney, H.A. Lubchenco, J. Melillo, J.M. Human domination of earth's ecosystems, Science, 277(5325), 1997, pp. 494-499.

Vitousek, P.M. Mooney, H.A. Lubchenco, J. Melillo, J.M. Human domination of earth's ecosystems, Science, 277(5325), 1997, pp. 494-499.

Westing, A.H. Establishment and Management of Transfrontier Reserves for Conflict Prevention and Confidence Building, Environment Conservation, 25(2), 1998, pp. 91-94.

Wildlife Institute of India. 1998. National wildlife database. Dehradun, India.

Zipkin, E.F. DeWan, A. Royle, A.J. Impacts of forest fragmentation on species richness: a hierarchical approach to community modeling, Journal of Applied Ecology, 46(4), 2009, pp. 815-822.

Associate Professor & HOD,
Department of Botany,
Dhing College, Dhing, Nagaon, Assam.
email : sanjeebkumarnath@gmail.com

20. An Impact of Green Revolution

Aiswarya k[1],
Meghavarshini [G2],
Rakshitha R[3]

Abstract

The green revolution refers to the development of high yielding plant varieties especially of where and rice, that increased food supplies in 1940s-60s and staved off widespread starvation in developing countries .The rapid increase in the human population, adding 1 billon people every 14 years, placed, and continues to place, heavy burden on the worlds ability to meet the consequential food demand .Around 1 billon people in the world are currently malnourished. The green revolution of today needs to significantly increase food supplies, but without harming the environment, causing a loss of biodiversity, or fostering high food prices that especially affect the poor.

Keywords : Green revolution, Environment, Agricultural, Nature.

Introduction

The green revolution was started in many countries around the world between the 1950s till the last 1906s. Many research technology transfer initiatives occurred around the world, which were geared toward increasing agricultural production. **Norman Borlaug** is called the father of the green revolution as he started the green revolution with his genetic testing he created a hybrid wheat plant that could resist fungus and diseases along with a high yield.

Objectives

To study an overview of "Green revolution"

Concept :

1. Short Term :

The main short aim behind this revolution is to address India's hunger crisis during the second five years plan.

2. Long Term :

The long term goal of this revolution is to do modernization of agricultural practice in rural areas. This will lead to modernization of rural development, industrial development, infrastructure, raw material etc.

3. Employment :

Another main objective after this revolution is to provide employment to both agricultural and industrial workers

4. Scientific Studies :

Another objectives is to produce strong plants which could withstand extreme climates and diseases

Green Revolution

The third revolution, also known as the green revolution. Was a time of technology transfer initiatives that resulted in a significant rise in agriculture production and crop yields. These changes in agricultural began in developed countries after World War II and spread globally till the late 1908s. In the late 1906s, farmers began incorporating new technology such as high yielding varieties of cereals , particularly dwarf wheat and rice, and the white spread use of chemical fertilizers (To produce their high yields , the new seeds require for more fertilizer than traditional varieties) , pesticides , and controlled irrigation . Agriculture also saw the adoption of newer methods of cultivation, including mechanization these changes were often implemented as package of practices meant to replace traditional agricultural technology.The green revolution refers to the spread of advances in agricultural technology that began in Mexico and which led to a significant increase in food production in the developing world. In addition to producing larger quantities of food, the green revolution was also beneficial because it made it possible to grow more crops on roughly the same amount of land with a similar amount of effort. This reduced production costs and also resulted in cheaper prices for food in the market.\The ability to grow more food on the same amount of land was also beneficial to the environment because it meant than less forest or natural land needed to be converted to farmland to produce more food. This is demonstrated by the fact that from 1961-2008 , As the human population increase by 100% the amount of forest and natural land converted to farm only increase by 10% . The natural land that is currently not needed for agricultural land is safe for the time being, And can be utilized by animals and plants for their natural habitat.

The three primary components of the green movement methodology were:

1. Using genetically enhanced seeds (high yielding variety seeds).

2. Double cropping on the acreage that is currently used.

3. The ongoing growth of agricultural regions

Advantages

Reduces Greenhouse Gas Emission :

As the high yield methodology influences the carbon cycles via the atmosphere, it vastly reduces greenhouse gas emission and emission-free environments.

Increase in Food Production :

It uses various technologies and results in an increase in food production. It is a choice from the conventional method of agricultural.

Low Food Prices :

The whole market relies on the demand and supply process. As the yields are continuous, they meet the demand, and the supply becomes easy. High-yields produce more food items and lower the food prices for all consumers globally.

Increases Afforestation :

As the demand for food increases, deforestation also increases. Thus introducing a green revolution meets food needs and increases afforestation.

Continuous Yield :

It also offers a constant yield of crops irrespective of seasons.

Disadvantages

Qulity of Soil :

It encounters and reduces soil quality because the repetitive usage of the same crops on the land results in soil nutrient depletion.

Health Problems :

Consuming foods produced using pesticides and fertilisers will significantly impact health-related issues.

Lack of Biodiversity :

Creates more significant exposure to food chain and leads to the loss of beneficial hereditary attributes produced in conventional farming.

Seed Sterility :

Introducing new technologies leads to the prevention of future crop growth by composing seeds from mature plants.

Monocropping :

Green revolution promotes monocropping patterns, which causes various problems and reduces the production of high yield crops.

Conclution

The major revolution was a major achievement for many development countries and gave them an unprecedented level of national food security. Green revolution has done a lot of positive things, saving the lives of Millions peoples and exponentially increasing the yield of food crops. But environment degradation makes revolution an overall insecurities, short term solution to the problem of food insecurity so more sustainable and environment friendly system of cultivation needs to Practiced.

Student (B.Com FA)
Department of Commerce,
Karpagam Academy of Higher Education, Coimbatore, India
Student (B.Com FA)
Department of commerce,
Karpagam Academy of Higher Education, Coimbatore, India
Student (B.Com FA)
Department of Commerce,
Karpagam Academy of Higher Education, Coimbatore, India

www.ingramcontent.com/pod-product-compliance
Lightning Source LLC
Chambersburg PA
CBHW050222270326
41914CB00003BA/533